D0848694
EX LIBRIS
Pace
University
New York · Westchester
PACE UNIVERSITY
OPPORTVNITAS
MCMVI
NEW YORK
NEW YORK & WESTCHESTER
PLEASANTVILLE/BRIARCLIFF
NEW YORK

# The Structure of World Energy Demand

# The Structure of World Energy Demand

Robert S. Pindyck

The MIT Press
Cambridge, Massachusetts, and London, England

This book was set in VIP Times Roman by DEKR Corporation, printed and bound by The Murray Printing Company in the United States of America.

**Library of Congress Cataloging in Publication Data**

Pindyck, Robert S
The structure of world energy demand.

Includes bibliographical references and index.
1. Petroleum industry and trade. 2. Energy consumption—Mathematical models. I. Title.
HD9560.4.P5 333.7 79-12565
ISBN 0-262-16074-9

מוקדש לנורית יקירתי

# Contents

# Preface

This book consolidates and summarizes the results of an econometric study of the characteristics of world energy demand. Since the resources available for this study were limited, the scope of the study also had to be limited. Attention was focused on particular sectors of energy use (primarily the residential and industrial sectors, although some work was also done on the transportation sector), models were designed primarily to elicit information about the long-run structure of demand, with only limited attention given to the dynamic adjustment of demand in response to changing prices or income, and most of the empirical estimation and hypothesis testing was done for a limited set of industrialized countries (although some models were also estimated using data for a broader set of both industrialized and less developed countries).

Despite these and other limitations, these results should contribute to our understanding of the structure of energy demand. In addition, they may help to resolve such issues as the extent to which energy demand in the long run is responsive to price changes, the possibilities for interfuel substitution, the substitutability of energy with other factors of industrial production, the impact of energy price changes on macroeconomic output, and the ways in which energy demand differs in the industrialized versus the less developed countries. These issues have been the subject of considerable debate in recent years and are important to the design of both energy and economic policy.

This study itself was conducted as part of a larger project at MIT to develop analytical models of the world oil market. Several of the sectoral demand models constructed here have become part of a large simulation model of the oil market; that larger model describes world oil demand, competitive oil supplies, and the production capacities of the OPEC cartel countries.

This study—and the larger World Oil Project—were generously funded by the RANN Division of the National Science Foundation, under a grant (no. GSF SIA075-00379) to the MIT Energy Laboratory in association with the Department of Economics and the Sloan School of Management. Additional funds were provided by the Center for Energy Policy Research of the Energy Laboratory. This study would not have been possible without this financial support, and the help of the NSF and the Center for Energy Policy Research is gratefully acknowledged.

This study relied extensively on the research assistance of a number of MIT graduate students and Energy Laboratory staff members. The estimation of our demand models required the assembling, organizing, and documenting of a large and comprehensive database. The data-gathering effort was begun by Ken Flamm and Dan Duboff, then graduate students in the Economics Department and Sloan School, respectively. The final organization and documentation of the database was carried out by Jacqueline Carson, a staff member of the Energy Laboratory. Kevin Lloyd and Eric Rosenfeld, Sloan School graduate students, and Jeff Ward, an Economics Department graduate student, devoted a considerable amount of time to setting up and carrying out the estimations of the translog and other models of residential and industrial demands. Ross Heide estimated the models of energy demand in the developing countries, and, as part of his Sloan School Master's thesis, constructed the models of energy demand in the transportation sector. Finally, Wayne Christian and Vinod Dar carried out much of the testing and validation of the models, and developed the computer framework for their integration into the simulation model of the world oil market. The help of all of these individuals is greatly appreciated.

The computational work for this study was performed on MIT's IBM 370 computer, using TROLL, a system for the estimation and simulation of econometric models. TROLL is maintained by the Center for Computational Research in Economics and Management Science in the Sloan School of Management, and members of the staff of that Center provided invaluable assistance in the use of the system.

A number of individuals provided advice and assistance during the course of this study, as well as comments on earlier drafts of this book and the working papers that preceded it. My thanks for this help go to F. Gerard Adams, M. A. Adelman, Ernst Berndt, Melvyn Fuss, James Griffin, Jerry Hausman, William Hogan, Henry Jacoby, Dale Jorgenson, Edwin Kuh, James Sweeney, and Leonard Waverman. Special thanks is due to V. Kerry Smith, of Resources for the Future, for his detailed review and extensive comments on an earlier draft of the book.

Finally, my appreciation to Deborah Caldwell for her considerable patience and perseverance in typing the various drafts of the manuscript.

# The Structure of World Energy Demand

# 1 Introduction

The dramatic increases in energy prices that occurred when the Organization of Petroleum Exporting Countries (OPEC) quadrupled the world price of oil in 1973–74 has had profound implications for the economies of all the industrialized countries. Other commodities have experienced rapid and substantial increases in price—the prices of bauxite and coffee, for example, both tripled in recent years, and the prices of grains and other agricultural products have experienced price fluctuations on the order of 300 or 400 percent over the years. Few people, however, would be as concerned about these events, or expect them to have anywhere near the impact on our standard of living that increases in the price of energy are likely to have. Higher energy prices have contributed to reduced economic growth in many countries, and in the long run may result in basic changes in lifestyles.

How fast energy prices rise in the future will depend in part on the rate at which conventional energy resources become scarcer and more difficult to find, in part on the rate of technological change that lowers the cost of nonconventional energy sources, in part on the behavior of the OPEC cartel, and in part on the domestic energy policies of various countries. Although the dramatic increases of 1973 to 1975 are not likely to continue, we should probably expect to see energy prices continue to rise in real terms, at least slowly, for the next few decades.

There is little doubt that past and future increases in energy prices will have a dampening effect on energy demand, as well as at least a temporary dampening effect on employment and economic growth. The questions that are of interest now are, first, to *what extent* will higher energy prices reduce energy demand, and second, will higher energy prices (combined with perhaps less energy use) necessarily mean reduced economic growth and a lower standard of living? As we will see, these questions are partly interrelated—both the extent to which prices affect demand *and* the effect of prices on our standard of living depend on the role that energy plays in the production of other goods and as a part of consumers' overall purchases of goods and services.

As can be seen in table 1.1, energy is indeed a major item in the consumption baskets of private households. Expenditures on energy in 1965 accounted for 9 percent of total household consumption expenditures in the U.S., and 7 percent to 9 percent in Japan and the European countries. These numbers stayed level over the next seven or eight years as real energy prices fell slowly, but energy consumption grew. By 1974 these energy expenditure shares had risen above their 1965 levels, largely because of the major increases in energy prices that occurred in 1973–74. Similarly, the cost of energy as a fraction of the total cost of industrial production in 1965 was 3 percent in the U.S. and 4 percent to 8 percent in Japan and the European countries. These numbers also stayed relatively stable over the next seven years as real prices fell, but production became more energy-intensive and began increasing in 1974.

The importance and impact of higher energy prices is only partly due to the relative magnitudes of energy expenditures. It is more a function of the particular characteristics of energy demand—the ability of consumers to use less energy directly and to shift their purchases of goods to those that require less energy, and the ability of manufacturers to produce their goods using less energy and instead more capital. A basic objective of this book is to obtain a better understanding of these characteristics.

## 1.1 The Impact of Higher Energy Prices

The conventional wisdom, as reflected in both popular opinion and the working assumptions often used for energy policy analysis, is that the price elasticity of the demand for energy is very small. (Elasticities in the range of 0.2 are often casually suggested as a basis for policy analysis.) The argument behind this conventional wisdom is that increases in energy prices tend to have a much greater impact on consumers and energy-using producers than do increases in the prices of other commodities because of the critical role that energy plays, both in the consumption basket and as a factor of production. The argument is made, for example, that consumers have very little flexibility to decrease their use of energy, or even to substitute between alternative fuels, while the consumption of food and other goods can be adjusted much more easily in response to changes in price.

This argument is probably valid in the short run. If energy prices suddenly increase, consumers cannot in the space of one or two years replace their cars with smaller, more fuel-efficient ones, replace their energy-consuming appliances (refrigerators, air conditioners, for example) with more energy-efficient ones, insulate their homes, and take other measures to significantly reduce their energy consumption. If the price of oil should suddenly rise while the price of natural gas remained fixed (and if supplies of natural gas continued to be available), it is not economical to quickly switch a home-heating system from oil to gas, so the potential for interfuel substitution is quite limited in the short run. Similarly, industrial users of energy cannot change their consumption patterns very much in the short run. Most capital equipment was designed to consume a certain amount of energy, so that capital and energy must be used together, that is, they are complementary inputs to production. Thus, when energy prices increase, producers, at least in the short run, do not have the flexibility to shift to more capital-intensive and less energy-intensive means of production. (Some shift to the use of more labor is possible, but this is costly in most industrialized countries where labor has become an increasingly expensive factor input.)

While this conventional wisdom is probably true for the short run, it may be far from true in the long run. Indeed, an important question facing us today is just how much flexibility there is in the use of energy in the long run. The answer to this question has important implications for the design of energy policy, and also for our assessment of the impact of higher energy prices on such macroeconomic variables as inflation, employment, and economic growth. If the household demand for energy is indeed sensitive to price in the long run, then eventually the impact of higher energy prices on consumers' budgets will be reduced as the quantities of energy consumed are reduced, and tax policies designed to reduce or limit household energy consumption would have a reasonable chance of success. Similarly, if energy and capital or labor are substitutable in the long run, and if the long-run price elasticity of industrial energy demand is sufficiently large, then increases in the price of energy will tend to drive up the cost of manufactured output by a smaller amount, and therefore have a smaller macroeconomic impact.

The long-run impact of changes in energy prices on energy use has an impact on energy prices themselves. The world price of oil is largely determined by the OPEC cartel, and changes in this price tend to drive changes in the prices of other fuels. Thus OPEC has the ability within limits to manipulate world energy prices. Of course, the prices actually faced by consumers will depend also on the taxes and/or price controls in individual countries, but these prices are still very much a function of OPEC's decisions with regard to the world price of oil. OPEC's ability to raise price, on the other hand, is to a considerable extent dependent on the price responsiveness of total energy demand, as well as the price responsiveness of non-OPEC energy supply. If in the long run energy demand is quite price responsive, then this means that OPEC cannot increase oil prices very much in the future (without incurring significant revenue losses). Thus our ability to predict energy prices in the long run (that is, to predict OPEC's pricing behavior) depends in part on our understanding of the long-run price and income elasticities of energy demand.

Another important question is the extent to which individual fuels can be substituted for each other in the long run. Over the next two or three decades reserves of oil and natural gas may be reduced considerably, so that the availability of moderately priced energy will depend in part on the ability of electric utilities and industrial consumers of energy to switch from these fuels to coal or perhaps nuclear power. In the somewhat shorter term, the impact on oil demand of increases in natural gas prices in the U.S. will depend on the extent to which these fuels are substitutes in different sectors. Finally, the extent of interfuel substitutability determines in part the impact of an increase in the price of oil or natural gas on the cost of manufactured output, and the impact of a shortage of oil or natural gas on the level of output. Thus a better understanding of the extent of interfuel substitutability and the magnitudes of cross-price elasticities of fuel demands is needed if we are to be able to evaluate the impact of changing fuel prices, and if we are to be able to design intelligently an effective energy policy.

The main focus of this study is to determine the characteristics of energy demand in the long run. We have had a rather poor understanding of the response of energy demand in the long run to changes in prices and income, and this has made it difficult to design energy and economic policies. By working with models of energy demand rather different from those that have been used before and by estimating these models using international data, we can better understand the long-run structure of energy demand and its relationship to economic growth.

## 1.2 The Structure of Energy Demand

The ratio of energy demand to GNP has been fairly constant for the U.S. over the period 1950 to 1974, and this has led some people to believe that a more or less fixed proportionality between energy use and total economic output must always hold. Such a belief, however, is completely unfounded. Until 1974, energy prices in the U.S. declined slowly but steadily in real terms, while recently we have experienced large increases in energy prices—increases which may significantly alter the amount of energy used per dollar of output. In addition, energy use per dollar of gross output has varied considerably across countries, and for many countries the ratio has changed significantly over time. This can be seen in figure 1.1, where we have plotted energy consumption per million dollars of gross domestic product (measured in constant 1970 U.S. dollars) for the U.S., Canada, the U.K., the Netherlands, France, and West Germany.[1] Note that the four European countries have energy/output ratios well below those of the U.S. and Canada, the ratio for the U.K. has declined somewhat over time, and that for the Netherlands has increased over time. There would certainly seem to be no magic number for an energy/GNP ratio.

Why do we observe these differences in energy/output ratios across countries, and why have the ratios increased in some countries, decreased in others, and remained more or less level in still others? An answer often given to these questions is that there are significant differences in lifestyles across countries which result in different needs for energy. Examples often cited include different sizes of cars driven because of basic inherent differences in tastes, or differences in the extent of home heating because of different habits. While tastes and habits may indeed differ across countries, and across time in any one country, this certainly does not provide

1. Energy use in that figure and elsewhere in this study is measured in teracalories (Tcals). 1 Tcal = $10^{12}$ calories = $3.97 \times 10^{9}$ Btu. Note that the thermal content of a barrel of oil is roughly $1.5 \times 10^{-3}$ Tcals.

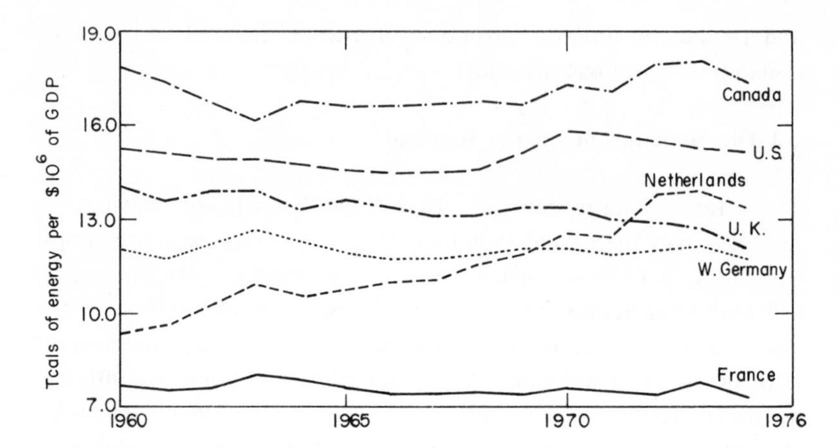


**Figure 1.1**
Energy use per million dollars of GDP (in 1970 U.S. dollars, converted using purchasing power parity indices for GDP)

a meaningful explanation for differences in energy use, and in particular does not provide a basis for predicting the kinds of changes in energy use that are likely to come about in the future from changing energy prices and changing GNP's. Tastes and habits may themselves be functions of price. Taking the price of gasoline and its relation to average car size as an example, we must ask to what extent car size differences can be attributed to differences in gasoline prices across countries and over time. It may well be that people choose to buy smaller and more fuel-efficient cars when gasoline prices are high (in fact get into the habit of driving smaller cars) for economic reasons. If this is indeed the case, it has important implications for the impact of higher gasoline prices (perhaps through taxes) on gasoline demand in the U.S. and elsewhere.

It is also important to recognize that differences in energy use cannot be explained in the aggregate but must be explained on a sector-by-sector basis. Clearly the structure of energy demand for home heating will be very different from that for industrial production, so that to look at energy/GNP ratios in the aggregate provides little in the way of useful information. Instead we must examine the characteristics of energy demand for each particular sector of use. In this study, we will focus primarily on three sectors of use: residential, industrial, and transportation.

For the residential or household sector, the structure of energy demand depends on consumers' preferences, and in particular the willingness of consumers to substitute between energy and other goods in their consumption baskets. For consumers, substituting away from energy as energy prices rise means less direct use of energy (for example, for home heating and cooling), as well as a reduction in purchases of energy-consuming appliances or the replacement of existing appliances with those that are more energy efficient. The characteristics of consumers' preferences also determine whether the consumption of energy (and the consumption of other items) rises proportionately with income growth, or whether income growth, with prices of all goods held fixed, is by itself likely to produce shifts in the proportions of expenditures allotted to energy and other categories of consumption.[2] Such shifts might occur, for example, if rising incomes encourage a more than proportional increase in energy-intensive consumption (through, say, increased purchases of labor-saving appliances). Finally, the extent to which fuels will be substitutable with each other as their relative prices change (given some overall level of energy use) is also likely to be different in the residential sector than in other sectors. In fact it will depend on the extent to which consumers prefer certain fuels for intrinsic qualities, such as cleanliness and security of supply, and in the long run the capital cost of substituting alternative fuel-burning appliances.

For the industrial sector, the structure of energy demand depends on the characteristics of production, and in particular the extent to which capital, labor, and energy can be used in different proportions in response to changes in the prices of these factors. The substitutability of capital, labor, and energy is a critical determinant of the industrial demand for energy, and also, as we will see, determines the macroeconomic impact of changes in energy prices. The characteristics of production further determine whether the industrial demand for energy, and the demands for other factors, rise

2. If shares of expenditures on particular consumption categories all remain fixed as income increases, then consumers' preferences are said to be "homothetic." We will discuss this concept in some detail in the next chapter.

proportionally with the growth of industrial output or whether output growth, with prices of factors held fixed, will by itself produce shifts in the proportions of expenditures allotted to each factor.[3] And finally, the characteristics of production determine the extent to which individual fuels will be substituted for each other as their relative prices change.

In the transportion sector, the demand for energy will depend on the demand for the specific form of transportation itself and the share in the cost of the transportation service represented by the cost of energy. (If energy costs are only a small share of the cost of the transportation service, then increases in the price of energy will only make a small change in the price of the service, and hence only have a small impact in the demand for the service even if that demand is highly elastic with respect to changes in the price of the service itself.) In addition, the demand for energy will depend on the ability to adapt the particular form of transportation to make it more fuel efficient (for example, by building smaller cars or driving existing cars at slower speeds). The demand for energy in the transportation sector will, of course, vary considerably across particular forms of transportation because of differences in the demands for the alternative transportation services themselves, differences in the energy cost shares for the services, and differences in the costs of improving fuel efficiency.

Energy is used in other sectors of the economy as well, in particular for energy transformation: largely the production of electricity, where interfuel substitution is an important aspect of demand, and as chemical feedstocks for the production of plastics and other basic materials, where interfuel substitution is less important. The scope of this study is limited, and we will touch only briefly on the characteristics of energy demand in these and other sectors.

### 1.3 Energy and the Macroeconomy

We will also be concerned in this study with the interrelationships between energy prices and energy use and such macroeconomic

3. The production structure is said to be homothetic if expenditure shares for each factor remain fixed as the total value of output increases. This, too, will be discussed in detail in the next chapter.

variables as employment, inflation, and GNP growth. It is important to recognize that the causal relationship between energy and the macroeconomy runs in both directions. Most people are aware of how increases in energy demand are brought about by growth in GNP, but only recently have people become aware of the importance of energy to GNP growth itself. A physical shortage of energy (or for that matter any other input used for production) can obviously depress GNP and increase unemployment by creating bottlenecks in the production of both intermediate and final goods. An increase in the price of energy, however, can also reduce the productive capacity of the economy. If energy or any other factor of production becomes more scarce (that is, more costly), this necessarily reduces the production possibilities of the economy, so that GNP will be lower than if energy prices had not increased. The question, of course, is how much lower GNP will be as a result of an increase in energy prices. Again this depends on the substitutability of energy with other factors. If the possibilities for substitution are great, then less expensive factors can be used in greater quantity in place of energy.

Because of the important interrelationships between the energy sector and the macroeconomy, energy use and energy policy are becoming increasingly important to the design of economic policy. We are beginning to realize, for example, that the rate of unemployment may depend not only on the particular monetary and fiscal policy in effect but also on changes in energy prices and energy use that took place over the last two or three decades.

To see this, consider the fact that between the end of World War II and 1972 a slow but steady shift occurred in the structure of industrial production in the U.S. and most of the other advanced economies. During this period two factor inputs of production—energy and capital—became significantly cheaper in real terms relative to a third important input, labor. This shift in relative prices occurred for a number of reasons. Reserves of energy resources, and energy production, were increasing worldwide, which drove down the real cost of energy. Tax policies in many countries (for example, the investment tax credit in the U.S.), designed to encourage new capital investment as a spur to economic expansion, helped to reduce the growth in the price of capital services. Finally, tax and social welfare policies, combined with greater wage demands on the part of workers, tended to greatly increase the effective cost of labor services for production. For the case of the U.S., this is illustrated in figure 1.2, which shows real price indices for capital, labor, and energy.

The result of these changing prices was a shift in the factor mix used in production. Gradually, producers replaced labor with less expensive capital and energy. In addition, there is evidence that capital and energy themselves came to be used in a complementary fashion. Since the particular forms of capital required large amounts of energy to be utilizable, there was little or no room for substitution between energy and capital, at least in the short run. (As we will see in this study, the evidence indicates that there may be more room for substitution in the long run when existing capital can be replaced by new, more energy-efficient, machines.) This shift in the relative quantities of factor inputs is illustrated in figure 1.3, which shows again for the U.S. quantity indices for capital, labor, and energy.

This shift away from labor and towards energy and capital served to exacerbate the impact of the increases in energy prices that were brought about by the OPEC cartel. When energy prices rose, industries in many countries were unable to achieve a significant shift away from energy-intensive production. For at least the short term,

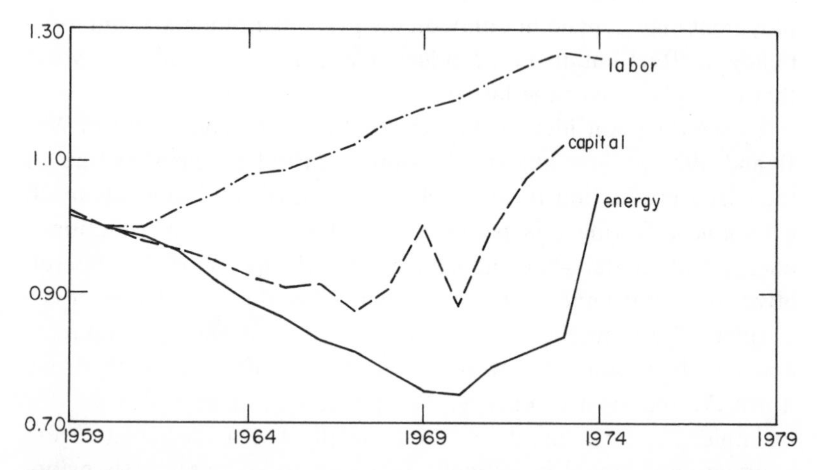


**Figure 1.2**
Real price indices for capital, labor, and energy in the U.S. (prices = 1 in 1960)

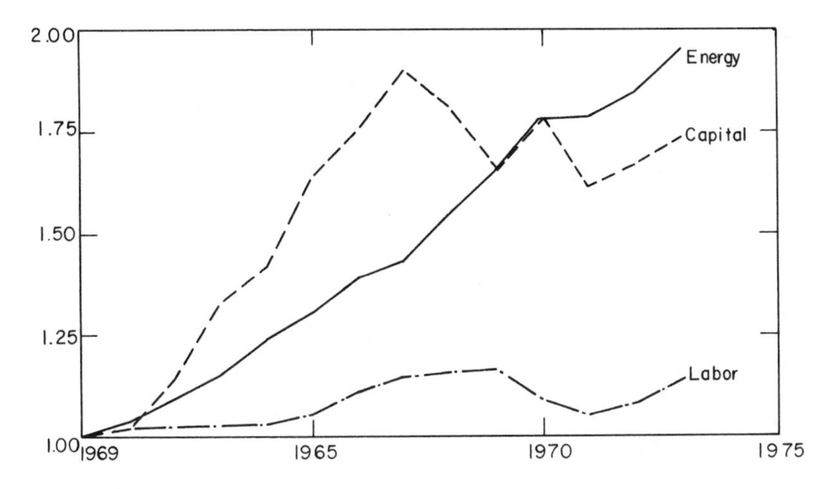


**Figure 1.3**
Quantity indices for capital, labor, and energy in the U.S. (quantities = 1 in 1960)

energy and capital were complementary inputs, and the only substitutable alternative—labor—was already very expensive. Thus increases in energy prices were translated into an increase in the cost of industrial output—an increase in cost nearly as large as the percentage increase in the price of energy times energy's share in the total cost of output. But with labor and capital fixed in the short run, this meant a drop in the level of real output. The result was a recession, and the likelihood of lower economic growth during the next several years.[4] In the short run at least, the shift towards more energy- and capital-intensive production meant a greater reduction in the productive capacity of the economy as a result of higher energy prices. The shift itself, on the other hand, came about because of gradual changes in energy prices (as well as changes in the prices of other factors).

Even if energy prices do not rise very rapidly in the future (and we will argue in chapter 8 that a fairly slow but steady rise in energy prices is in fact a more probable scenario for the future), the large and dramatic increase in energy prices that has already occurred will cause some reduction in economic growth in the industrialized

4. We will discuss this problem later in chapter 8.

countries, at least through 1985. The question now is the extent to which growth will be diminished over this intermediate range. The answer depends in part on the degree of substitutability of capital, labor, and energy in the long run. If capital and energy are substitutable, then the impact of higher energy prices on the cost of output (and thus on GNP) will be ameliorated. Determining the long-run relationship between capital, labor, and energy in aggregate production is therefore an important objective of this study.

### 1.4 The Need for an International Study

An important feature of this study is that it deals with differences in energy prices and energy use across a number of industrialized countries. There are three basic reasons for conducting an international study of energy demand. First, the use of international data permits us to identify the long-run structure of energy demand to an extent not possible using data for a single country. Second, we are interested in some of the ways in which the structure of energy demand might differ across countries, as well as the possible reasons for such differences. Third, we would like to better understand how world energy markets, and the world demand for energy, are likely to change in response to changes in the prices and availabilities of energy supplies.

As we explained above, our objective in this study is to analyze the long-run characteristics of energy demand. This is difficult to do using data for a single country. Until recently energy prices in most countries have changed only slowly over time, so that the estimation of models of energy demand for a single country is most likely to capture short-run or intermediate-run price and income elasticities. In order to estimate long-run elasticities of demand, it is necessary to compare the equilibrium demands for energy corresponding to prices that are significantly different from each other. By equilibrium demand we mean the demand that would prevail after sufficient time had elapsed for consumers to completely adapt to a new price or set of prices. How much time is sufficient will depend on the particular sector (or even subsector) of energy use, but it might be anywhere from five to twenty years.

Given the limited time horizon for which data is available for any one country, it is clear that we cannot compare equilibrium prices

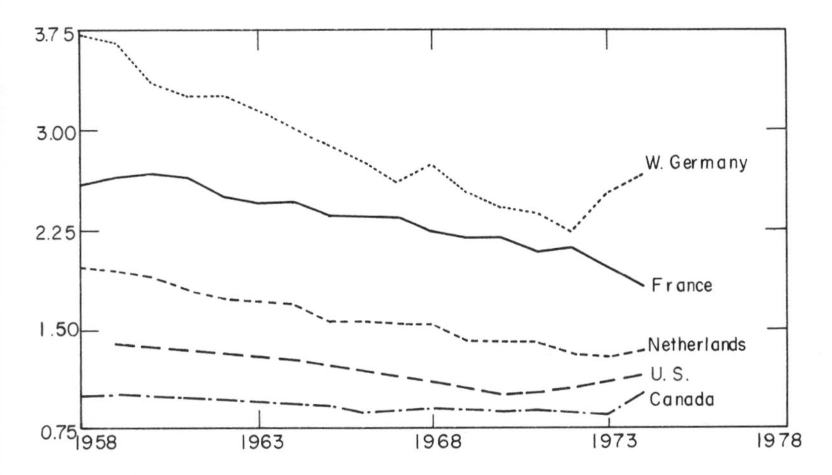


**Figure 1.4**
Energy price index for the residential sector (price = 1 for the U.S. in 1970)

and demands by working with time-series data for only a single country.[5] Cross-sectional data might be used that span a number of different regions in a single country (for example, interstate data for the U.S.), but in most cases energy prices and per capita consumption levels show little regional variation within a country. On the other hand, energy prices are and have been quite different across countries, so that by using data that span a number of countries, we can effectively compare long-run equilibrium values of energy prices and demands.

The variation of prices across countries and through time is illustrated for the residential sector in figures 1.4 to 1.6. Figure 1.4 shows a real price index for energy (all prices relative to a price of 1.0 in the U.S. in 1970) for the U.S., the Netherlands, West Germany, Canada, and France. (The computation of this index is discussed in chapters 3 and 5.) Note that these prices have declined over time in all countries, but only slowly. The prices, however, vary considerably across countries, with energy prices in West Germany (the highest) about triple those in Canada (the lowest).

5. As we will argue later, it is for this reason that most econometric work on energy demand based on a single country has yielded price elasticities that are quite low—they are probably short-run and not long-run elasticities.

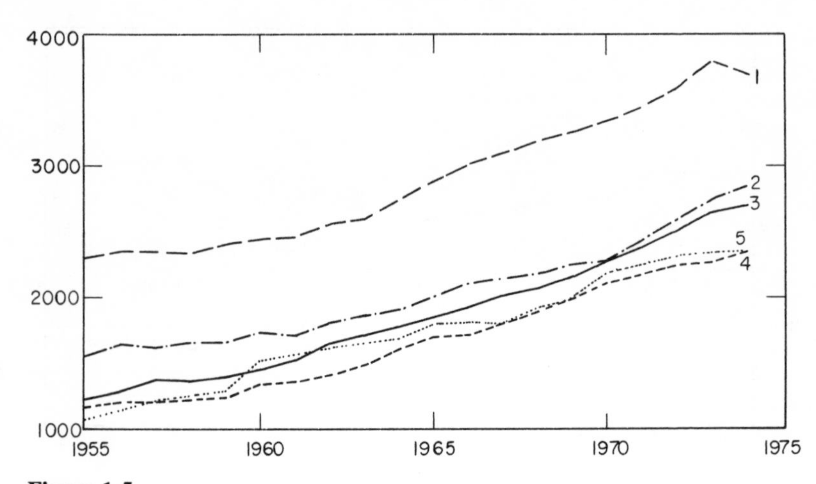


**Figure 1.5**
Per capita incomes (all in 1970 U.S. dollars): (1) U.S., (2) Canada, (3) France, (4) Netherlands, (5) West Germany

Figure 1.5 shows per capita disposable incomes in these same five countries, and again there is considerable variation across countries, but only slow change over time. Finally, per capita energy consumption in the residential sectors of each of these countries is shown in figure 1.6. These levels of energy consumption vary slowly over time, but the greatest variation is across countries, so that they can be viewed as comparative equilibrium levels.

There are certain problems involved in pooling data across a number of countries, the most serious of which is the possibility that the fundamental structure of demand may differ from one country to the next. In the industrial sector, for example, if the underlying structure of production (as represented in our models by a production function or cost function) differs across countries in a pooled sample, biased elasticity estimates may result. We try to minimize this potential hazard by testing wherever possible the homogeneity of demand structures across countries. This in turn provides information on the extent of and reasons for intrinsic intercountry differences in the structure of demand—which brings us to our second reason for conducting an international study such as this one.

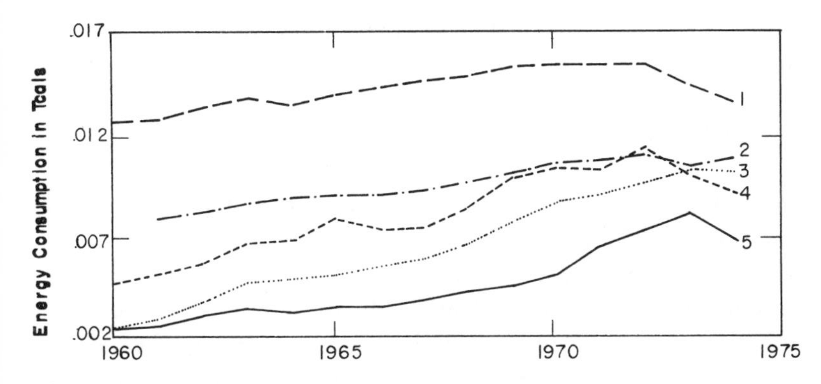


**Figure 1.6**
Per capita residential energy consumption: (1) U.S., (2) Canada, (3) West Germany, (4) Netherlands, (5) France

We wish to use the considerable differences in energy prices and energy demand levels across countries to obtain estimates of long-run demand elasticities, but also to determine if and how these elasticities might differ across countries, or across groups of countries. For example, we would expect long-run price and income elasticities to differ considerably between developed and less developed countries, but we also wish to determine whether these elasticities differ across the developed countries. One way to find this out is by estimating energy demand models using alternative subgroupings of countries, so that each subgroup displays enough cross-country variation in the data to identify long-run elasticities of demand. We will use such an approach to determine whether demand elasticities in, say, Canada and the U.S. differ from those in some of the European countries.

The third reason for conducting an international study of energy demand is to get a better understanding of how world energy markets are likely to evolve in the future and how the total world demand for energy is likely to change in response to the increases in oil prices initiated by the OPEC cartel, and in response to future changes in prices resulting from the cartel or from changing supplies in noncartel countries. This is important since it is a determinant of what OPEC itself can do to the price of oil in the future and, combined with projections of future energy production capacities, helps us determine the likelihood of stability in energy markets and energy supplies.

## 1.5 The Plan of the Book

This book describes the results of an econometric study of the world demand for energy. An important objective of the study has been to develop and estimate models that can better reveal the structure of energy demand and help to determine the long-run response of demand to changes in prices and levels of economic activity. We have concentrated on the residential, industrial, and transportation sectors of several industrialized countries, although we also briefly consider the demand for energy in other sectors, and in some of the developing countries. Finally, the study has been concerned with the relationship between energy use and economic growth, and in particular the impact of higher energy prices on the cost and level of economic output.

Our approach has been to specify and estimate consistent models of energy demand. By "consistent" we mean models that simultaneously describe the demands for energy and other competing commodities, as well as the demands for individual fuels within the energy aggregate. In studying the industrial sector, for example, we do not view energy in isolation but rather as one of three inputs to production, and therefore construct a model that simultaneously accounts for the demands for capital, labor, and energy. Similarly, the price of energy is derived from a submodel that describes the interrelated demands for each of four fuels (coal, oil, natural gas, and electricity).

Usually a model for the demand of some commodity presupposes a particular demand structure. The problem with this is that the estimated elasticities from this model are valid only insofar as the specification itself is valid. Our approach is to use functional specifications for our models that are as general as possible, in other words, that impose no or almost no a priori restrictions on the structure of demand. We then estimate these models, using pooled international time-series cross-section data in order to test particular restrictions on the structure of demand and to identify long-run demand elasticities.

The specification of our demand models is explained in some detail in the next chapter of this book. There we discuss the use of the indirect translog utility function as a basis for modelling the residential demand for energy, and the use of the translog cost function as a basis for modelling the industrial demand for energy. In addition, we introduce alternative demand specifications for the residential and industrial sectors, and also discuss model specifications for the transportation sector and for other sectors.

There are a number of important methodological issues that must be dealt with in estimating models of energy demand, and these are discussed in chapter 3. These issues include the use of purchasing power parities to convert prices and expenditure figures in the local currencies of various countries to common units, the measurement of energy use in ''gross'' versus ''net'' (efficiency-adjusted) terms, the calculation of aggregate energy price indices, the estimation of intercountry differences in energy demand, and the choice of econometric technique. In addition, chapter 3 discusses the construction of all the data series used in the estimation work.

The statistical results of this study are presented in chapters 4, 5, and 6. Chapter 4 deals with the residential sector, and discusses the role of energy as part of total consumption expenditures. This chapter presents alternative elasticity estimates for total energy use in the residential sector, as well as estimates describing the substitutability of fuels used in that sector. Chapter 5 deals with the industrial sector, and describes estimates of the elasticities of substitution between energy, capital, and labor, elasticities of total energy demand, as well as own- and cross-price elasticities for individual fuels. Chapter 5 also deals with the impact of higher energy prices on the cost of industrial output, and provides estimates useful for analyzing the macroeconomic impact of changing energy prices. Finally, chapter 6 describes the estimation of a model of motor gasoline demand, as well as elasticity estimates obtained for other fuels used in the transportation sector.

Chapter 7 deals with energy demand in the less developed countries. Since meaningful price and quantity data is very difficult to obtain for these countries, our statistical work here is only very limited. However, we offer some arguments about the characteristics of energy demand in these countries, test these arguments to the extent possible given the data, and speculate on the ways in which demand might change in the future.

Finally, chapter 8 deals with the macroeconomic impact of higher energy prices in the industrialized countries, and the likely future evolution of world energy markets. In particular, this chapter considers the effects of changing energy prices on economic growth in the industrialized countries, the likelihood of increases in energy prices in the future, and the impact of price changes on world energy demand and supply. The chapter concludes with some remarks about the implications of our results for energy policy.

### 1.6 A Word on Units of Measurement

Energy can be measured using any one of a plethora of different units—calories, Btu, barrels of oil equivalent, and million tonnes of oil equivalent to name just a few—and this often creates confusion for readers of articles and books on energy. In this book we try to measure all energy quantities in teracalories (1 Tcal = $10^{12}$ calories). In some cases, however, it is more natural to refer to other units (such as dollars per gallon for the price of gasoline). In addition, many readers may want to convert our numbers into units they are more familiar with. We therefore present a set of useful conversion factors in table 1.2.

Note that there are basically three ways of measuring a quantity of energy: as a physical quantity of a particular fuel (such as tons of coal or barrels of oil), in terms of thermal content (such as millions of Btu), or in terms of the thermal-equivalent quantity of some numeraire fuel (such as barrels of oil-equivalent). The first approach is straightforward, and as long as the particular fuel in question is specified exactly (for example, coal of a particular grade and thermal content), there is no measurement error involved—a cubic foot of high Btu gas is a cubic foot of high Btu gas. However, this approach is not useful if we wish to talk about aggregate energy use, and add up coal, oil, and so forth. For this reason energy quantities are often measured in terms of thermal content or the thermal-equivalent quantity of a numeraire fuel.

Unfortunately these last two ways of measuring energy quantities can lead to certain errors. The problem is that the thermal content of a fuel depends on the particular way the fuel is burned, and fuels can be burned in very different ways. Although the conversion to thermal-content units is usually done assuming 100 percent thermal efficiency (that is, complete burning), different fuels and different uses of the same fuel can involve different thermal efficiencies, and this can make quantity comparisons misleading. Here we simply raise this issue as a warning; it is discussed in more detail in chapter 3.

**Table 1.1**
Magnitudes of energy expenditures

| | Energy expenditures by households as fraction of total household consumption expenditures[a] | | Energy expenditures by industry as fraction of cost of industrial output | |
|---|---|---|---|---|
| | 1965 | 1974 | 1965 | 1973 |
| U.S. | 0.087 | 0.100 | 0.029 | 0.033 |
| U.K. | 0.083 | 0.084 | 0.069 | 0.047 |
| France | 0.068 | 0.076 | 0.052 | 0.045 |
| West Germany | 0.072 | 0.125 | 0.047 | 0.045 |
| Italy | 0.068 | 0.073 | 0.074 | 0.066 |

[a] includes motor gasoline

**Table 1.2**
Units of measurement

**A. Physical quantities**

1 metric ton (tonne) = 2,204 lbs = 0.984 long tons = 1.102 short tons = 1,000 kg

1 barrel crude oil = 42 U.S. gallons, and weighs 0.136 metric tons

1,000 cu ft (1 mcf) natural gas = 28.3 cu meters

1 kilowatt-hour (kwh) of electricity = 3411 Btu = 860 kilocalories (kcals)

1 Btu of energy = 252 calories = $10^{-5}$ therm

1 Tcal of energy = $10^{12}$ calories = $4 \times 10^9$ Btu

1 Quad of energy = $10^{15}$ Btu = $2.5 \times 10^5$ Tcals

**B. Thermal equivalents**

1 metric ton anthracite coal = $2.80 \times 10^7$ Btu = $7.06 \times 10^9$ calories

1 metric ton bituminous coal = $2.89 \times 10^7$ Btu = $7.28 \times 10^9$ calories

1 barrel crude oil = $5.80 \times 10^6$ Btu = $1.46 \times 10^9$ calories

1 barrel residual fuel oil = $6.29 \times 10^6$ Btu = $1.58 \times 10^9$ calories

1 barrel distillate fuel oil = $5.83 \times 10^6$ Btu = $1.47 \times 10^9$ calories

1 barrel regular gasoline = $5.25 \times 10^6$ Btu = $1.32 \times 10^9$ calories

1,000 cu ft natural gas = $1.035 \times 10^6$ Btu = $2.61 \times 10^8$ calories

**C. Thermal-equivalent quantities of oil**

1 metric ton of coal $\simeq$ 4.9 barrels of crude oil

1,000 cu ft natural gas = 0.178 barrels of crude oil

1,000 kilowatt-hours electricity = 0.588 barrels of crude oil

$10^6$ Tcals of energy = 93.8 million tons of oil equiv. (Mtoe) = 1.89 million barrels per day (mb/d)

1 Quad of energy = 23.5 million tons of oil equiv. = 0.474 mb/d

# 2 Models of Sectoral Energy Demand

The demand for energy has been the focus of a growing number of studies, but the evidence on the structure of energy demand and the long-run response in energy demand to changing prices and changing levels of economic activity is still mixed. Some recent statistical studies demonstrate that there exist widely conflicting estimates of price elasticities of aggregate energy use in different countries and of own- and cross-price elasticities of the demands for individual fuels.[1]

This wide variation in energy demand elasticity estimates has occurred for several reasons. First, the models used to describe and characterize energy demand have themselves varied widely. Many of these models have imposed important a priori restrictions on the structure of energy demand—restrictions that may or may not be valid. For example, numerous early studies of the demands for particular fuels (as well as the demands for many other goods) use double logarithmic demand functions with or without time trends to allow for autonomous shifts in demand. The double logarithmic demand function is convenient in that price and income elasticities are constant and given directly by the estimated coefficients of the price and income variables. This function, however, implies some severe restrictions on the structure of demand, and cannot be derived as an exact representation of any well-behaved utility function. If in fact this restricted representation of demand is not a correct specification of consumer behavior, the elasticities resulting from its estimation may be biased.

A second problem, and one that is related to the first, is that many earlier studies model the demand for a particular fuel without considering its relationship to other fuels, or model the demand for total energy use without considering its relationship to the other components of consumer expenditures in the residential sector or to other factor inputs to production in the industrial sector. In the residential sector, for example, consumers might demand energy for heating or other purposes but may choose among alternative fuels in satisfying that energy demand. The actual shares of energy expenditures allotted to particular fuels will, of course, depend on relative fuel prices and perhaps other variables, but clearly these

1. See, for example, the recent surveys by the U.K. Department of Energy (16), Taylor (155), or those at the ends of chapters 4, 5, and 6.

shares must sum to 1. Often models of the demands for individual fuels (the double logarithmic model is an example) are additively inconsistent, in that predicted fuel expenditures need not add up to total energy expenditures. This implies a structural misspecification which again may result in biased elasticity estimates.

A third problem with many models is that they do not properly capture the long-run structure of demand, and mistake short-run elasticities for long-run ones. This problem arises because these models are often estimated using time-series data that pertain to a particular country. Such data usually elicits only a short-run response to price and income changes. Even if the specification of such a model is structurally dynamic (that is, contains lag relationships that differentiate between short-run and long-run responses) the use of pure time-series data may result in parameter estimates such that the predicted long-run elasticities really pertain to the short or intermediate run. Such models could yield a whole range of long-run elasticity estimates that are well below the true long-run elasticities.

A fourth problem has to do with sectoral and/or regional aggregation. The structure of energy demand is quite different for different sectors of use, so that the meaningfulness of models that describe total energy use for the entire economy is likely to be rather limited. Similar difficulties can arise from estimating models with regionally aggregated data. Studies of natural gas demand in the U.S., for example, show clear variation in price and income elasticities across different regions.[2] Unfortunately, data limitations often make regional disaggregation impractical. In this study we work with a number of different countries, and while separate models are developed for each sector of energy use, we cannot regionally disaggregate the individual countries themselves.

In this study we have tried to surmount these problems as much as possible. We succeeded in doing this for some sectors of energy use in some countries, but limited time and resources made it impossible to deal with these problems for all sectors or for all countries. For the residential and industrial sectors we were able to develop fairly detailed models of energy use for the major energy-consuming countries. The approach in these cases was to work with

2. See P. W. MacAvoy and R. S. Pindyck (113) and R. S. Pindyck (137).

generalized functional forms that impose no a priori restrictions on the structure of demand but instead allow particular restrictions to be tested empirically. In the residential sector this meant working with the indirect translog utility function, and for the industrial sector, the translog cost function. (These functional forms also yield models that are consistent in that expenditure shares sum to 1.) In addition, we work with pooled time-series cross-section data as a means of identifying long-run elasticities. However, for other sectors of use, and for a number of countries that are not major consumers of energy, we were forced to work with much simpler models of demand, models that have many of the problems we have mentioned.

In this chapter we discuss the theoretical specifications of our models of sectoral energy demands. For the residential sector we have concentrated on the indirect translog utility function as a means of describing the breakdown of consumer expenditures into energy and several other categories, and the breakdown of energy expenditures into expenditures on individual fuels. For the industrial sector, we discuss the use of the translog cost function to describe the allocation of production costs to three factors of production—capital, labor, and energy—and to describe the breakdown of energy expenditures into expenditures on fuels. For the transportation sector, we discuss a simultaneous-equation model of the demand for motor gasoline that describes the stock of cars, the use of that stock, and the average gas-burning efficiency of the stock. Energy demand in the remainder of the transportation sector, and in other sectors, is described using double logarithmic equations and other restrictive functional forms.

## 2.1 The Residential Sector

The residential sector accounts for an important component of total energy use and has been the target of energy conservation measures in a number of countries. Import-dependent countries in particular have imposed stiff taxes on fuels to limit energy consumption or to change the fuel mix in aggregate energy use. However, conservation policies are often designed in the absence of any reliable estimates of their probable long-run impact on energy use. The effects of rising per capita incomes and changing fuel prices on household energy demand depend on the willingness of consumers to substitute energy and energy-consuming goods for other consumption items, and to substitute among fuels within the energy aggregate. The household demand for energy has been the focus of a number of recent studies, but the evidence on the role of energy in the consumption basket and on interfuel substitution is still mixed.[3]

Our approach in modelling the residential demand for energy is to assume that consumers make two simultaneous utility-maximizing decisions in purchasing fuels. With the amount of money to be spent on energy taken as given, consumers allocate these expenditures among fuels—oil, natural gas, coal, and electricity. Consumers further decide what fraction of their total consumption budget will be spent on energy as opposed to other categories of consumption such as food and clothing. This approach requires the implicit assumption that consumers' utility functions are groupwise separable between fuels and other commodities, such that expenditure shares for fuels may depend on total energy expenditures but are independent of the expenditure shares for other consumption categories.[4]

Energy demand in the residential sector, and in particular the characteristics of interfuel substitution, will be highly dependent on the stock of energy-using appliances, at least in the short run. When fuel prices change, consumers cannot quickly switch heating systems or shift to more fuel-efficient appliances. Our model, however, does not treat energy as a (partially) derived demand from an appliance stock—in fact, in our model durable goods are a separate consumption category. One reason for this is that a derived demand model requires a detailed model of the appliance stock disaggregated by fuel type, and the required appliance stock data is simply not available for an international study such as this.[5] A second reason

3. We survey the results of some of these recent studies and compare them with our own at the end of chapter 4.

4. We thus have a utility tree along the lines described by Strotz (151), and the marginal rate of substitution between any two variables in the class of energy expenditures is independent of the expenditures on any other consumption category.

5. For an example of a derived demand model of interfuel substitution, see Pindyck (133). That model was specified for the U.S., but even there sufficient data was not available for estimation.

is that we are primarily interested in the long-run structure of demand, and although appliance stocks are an important determinant of both short-run energy demand and the speed of adjustment of demand, they play a much smaller role in the determination of long-run demand.

For the first stage of our two-stage model of residential energy demand we need a model for the breakdown of total consumption expenditures into energy and other categories. A number of such models have been constructed by others, some of them additively consistent (such that expenditure shares add to 1) and some inconsistent.[6] Typical model choices have included the additive logarithmic model, the linear expenditure system, and the additive quadratic model.[7] Usually these models have been estimated using time-series data for single countries, but in some cases cross-country comparisons have been made using pooled time-series cross-section data for a number of countries.[8]

The problem with these models, however, is that they all imply serious restrictions on the underlying structure of consumers' preferences: we would like to use a functional specification that is as general as possible, so that particular restrictions can be tested empirically rather than assumed a priori. We therefore work with various versions of the indirect translog utility function. The advantage of this functional form is that it is a general approximation to any indirect utility function, and therefore does not a priori impose constraints of homotheticity or additivity. Instead such constraints can be tested, as can the hypothesis of utility maximization itself.

In describing the demand for individual fuels, we will also work with alternative versions of the indirect translog utility function. In so doing we can obtain unrestricted estimates of own-price, cross-price, and total expenditure elasticities for each fuel, and then test for particular restrictions on the structure of demand. Although we have found the indirect translog utility function to be preferable, we

6. For an overview of models of consumer behavior see Brown and Deaton (25) and Phlips (132).
7. See Houthakker (89), Pollak and Wales (142), Houthakker and Taylor (92), Phlips (131), and Theil (158).
8. See Houthakker (91), and Goldberger and Gameletsos (66).

will also report on the use of alternative models of interfuel substitution, including static and dynamic multinomial logit models of consumer choice.

Our models for the residential sector are estimated using pooled time-series cross-section data for nine countries: Belgium, Canada, France, Italy, the Netherlands, Norway, the U.K., the U.S., and West Germany. Pooling countries is essential since our objective is to identify long-run elasticities. However, we also estimate these models using alternative poolings in order to identify possible cross-country differences in elasticities. In addition, we work with a nonstationary version of the indirect translog utility function so that long-run price effects can be isolated from changes in demand resulting from changing tastes.

As one might expect, work such as this is continually bound by data limitations. For many countries there are no good data available for some or all of the variables of interest to us, while for other countries data exists, but obtaining that data can be an extremely time-consuming and laborious task. As a result choices had to be made as to which data were to be collected. These data limitations were one of the factors that helped define and delimit the modelling approaches used here. In particular, it necessitated restricting our detailed analysis of residential demand to the particular set of countries just mentioned. Even for these countries, however, the quality of the data varies, and compromises in model specification and estimation had to occasionally be made.

**Use of the Indirect Translog Utility Function**

Our models of residential energy demand are derived from the underlying structure of consumers' preferences as represented by their indirect utility functions. The indirect utility function is the dual of the utility function $u = u(x_1, x_2, \ldots, x_n)$, where the $x_i$ are the consumption levels of the various goods or categories of goods. Given a set of prices $P_1, \ldots, P_n$ and an income constraint $M$, utility maximization leads to a system of demand equations:

$$x_i^* = x_i^*(P_1, \ldots, P_n, M), \qquad i = 1, \ldots, n.$$

Replacing $x_i$ by the optimal $x_i^*$ in the direct utility function $u$ gives the indirect utility function $V$:

$$V = u^*[x_i^*(P_1, \ldots, P_n, M), \ldots, x_n^*(P_1, \ldots, P_n, M)]$$
$$= u^*(P_1, \ldots, P_n, M).$$

Note that the indirect utility function has prices and income as arguments and is homogeneous of degree zero in these arguments; since a proportional change in all prices and income does not affect the optimal consumption levels $x_i^*$, it cannot affect $u^*$ either.

The duality between the direct and indirect utility functions leads to an important relationship known as Roy's identity:[9]

$$x_i^* = -\frac{\partial V/\partial P_i}{\partial V/\partial M}.$$

We will be interested in the expenditure shares $S_i = P_i x_i/M$, for which Roy's identity becomes

$$S_i = -\frac{\partial \log V/\partial \log P_i}{\partial \log V/\partial \log M}. \qquad (2.1)$$

So far we have assumed that the functional form of the indirect utility function is completely general, that is, there are no particular restrictions regarding its structure. Certain restrictions may in fact apply to the indirect (or direct) utility function, and these include stationarity, groupwise separability, homotheticity, and additivity. Stationarity means that preferences do not change with time, so that $V$ is not an explicit function of time. Groupwise separability allows us to partition the set of consumption goods into subsets, which would include goods that are closer substitutes or complements to each other than to members of other subsets. In particular, groupwise separability implies that the marginal rate of substitution between any two goods in one subset is independent of the consumption of any good in another subset. Homotheticity of the indirect utility function means that the expenditure shares $S_i$ are independent of total expenditures $M$, and this implies that the income elasticities of demand for every good are the same and equal to unity. Finally, additivity of the indirect utility function means that it can be written as a sum of functions of each good. As Houthakker (90) has shown, additivity implies that all cross-price

9. See Roy (146), or, for a simple derivation, Phlips (132). For a detailed overview of consumer demand theory, see Phlips (132) or Theil (158).

elasticities with respect to a particular price are the same. If the indirect utility function is both homothetic and additive, then the demand functions have unitary own-price elasticities and zero cross-price elasticities.[10]

Usually the theoretical derivation of demand models involves first specifying a particular functional form for the direct or indirect utility function and then determining the demand relationships implied by utility maximization. Often, however, the underlying utility function implies one or more of the restrictions mentioned above. The linear expenditure system, for example, which has been used widely to model the breakdown of total consumption expenditures, is stationary and additive, although it is not homothetic.

In order to avoid imposing a priori restrictions on the structure of demand, it is desirable to work with as general a utility function as possible, and we therefore use the indirect translog utility function. The indirect translog utility function is a second-order approximation to any indirect utility function. With time-varying preferences, as introduced by Jorgenson and Lau (101), it has the form:[11]

$$\log V = \alpha_0 + \sum_i \alpha_i \log (P_i/M) + \alpha_t t + \frac{1}{2} \sum_i \sum_j \beta_{ij} \log (P_i/M) \log (P_j/M) + \sum_i \beta_{it} \log (P_i/M) \cdot t + \frac{1}{2} \beta_{tt} \cdot t^2. \tag{2.2}$$

When the indirect translog function is used to model expenditure shares for energy and nonenergy consumption categories, $P_i$ is the price index for consumption category $i$, and $M$ is total consumption

10. For a full discussion of these restrictions, see Phlips (132) or Theil (158).

11. The indirect translog utility function without time was introduced by Christensen, Jorgenson, and Lau (42). The homothetic form of the indirect translog function was used by Christensen and Manser (43) to study consumer preferences for food, and the nonhomothetic form was used by Jorgenson (98) to study a three-category breakdown of consumer expenditures in the U.S. Berndt, Darrough, and Diewert (17) demonstrated empirically that the translog specification is more robust than other generalized functional forms such as the generalized Leontief or generalized Cobb-Douglas utility functions.

expenditures. When this function is used to model fuel shares, $P_i$ is the price of fuel $i$, and $M$ is total expenditures on energy.

Under the assumption of utility maximization we can apply Roy's identity as expressed by equation (2.1) to obtain the budget-share equations:

$$S_j = \frac{P_j X_j}{M}$$

$$= \frac{\alpha_j + \sum_i \beta_{ji} \log (P_i/M) + \beta_{jt} \cdot t}{\alpha_M + \sum_i \beta_{Mi} \log (P_i/M) + \beta_{Mt} \cdot t}, \qquad j = 1, \ldots, n, \qquad (2.3)$$

where $X_j$ is the quantity consumed of category $i$ (or fuel $i$), $t$ is a time trend (equal to zero at the beginning of the estimation period), and

$$\alpha_M = \sum_k \alpha_k, \qquad \beta_{Mi} = \sum_k \beta_{ki}, \qquad \beta_{Mt} = \sum_k \beta_{kt}.$$

Note that the parameters $\alpha_0$, $\alpha_t$, and $\beta_{tt}$ in equation (2.2) do not affect the utility-maximizing quantities consumed, and therefore cannot be identified. In addition, note that the budget constraint implies that $\Sigma S_j = 1$, so that only $(n - 1)$ of the $n$-share equations need be estimated to determine all of the parameters.

Consistent estimation of these share equations requires the exogeneity of the prices $P_i$. It is indeed reasonable to assume that these prices are exogenous—which is one of the reasons we work with an indirect rather than direct utility function. (In the direct translog utility function the share equations are functions of the quantities of each consumption category, and it is less reasonable to assume that these quantities are exogenous.)

The budget-share equations must be homogeneous of degree zero in the parameters since a proportional increase in income and all prices should leave the share allocations unchanged. For this reason a parameter normalization is required for estimation. We use the convenient normalization $\alpha_M = \Sigma\alpha_k = -1$. A number of parameter restrictions are also required if the share equations are indeed based on utility maximization. In particular, the parameters $\beta_{Mi}$ and $\beta_{Mt}$ must have the same values in each of the $n$-estimated share equations. Since there are $(n + 1)$ parameters involved, and $(n - 1)$ equations are estimated, this implies a total of $\frac{1}{2}(n + 1)(n - 2)$ restrictions. Also we assume that the indirect utility function is well behaved, and in particular that log $V$ is twice differentiable in its arguments. This means that the Hessian of log $V$ must be symmetric, which implies the following $\frac{1}{2}(n - 1)(n - 2)$ symmetry restrictions:

$$\beta_{ij} = \beta_{ji}, \qquad i \neq j, \qquad i, j = 1, \ldots, n.$$

There are an additional $(n - 1)$ restrictions resulting from the fact that the parameters of the $n$th equation are determined from the parameters of the first $(n - 1)$ equations and the definitions of $\beta_{Mi}$ and $\beta_{Mt}$. Thus the total number of parameter restrictions is $\frac{1}{2}n(n - 1)$.

There are other restrictions that might be imposed on the indirect translog function, and statistical tests can be performed to determine whether such restrictions are supported by the data. We will test some of these restrictions in this work, so we list them here.

The indirect translog function is stationary if preferences do not change with time, and stationarity implies that the parameters $\beta_{jt}$ are all equal to zero, $j = 1, \ldots, n$.[12] It is important, however, that a stationary function not be confused with a static function. As we will see later, a dynamic translog function—in which long-run elasticities differ from short-run elasticities—may still be stationary as long as the elasticities themselves do not depend on the particular time in which prices or income change.

In estimating the consumption breakdown model, we can test for groupwise separability between energy and the other consumption categories. Letting $P_1$ and $S_1$ be the price index and expenditure share for energy, and $P_2, P_3, \ldots, P_n$ and $S_2, S_3, \ldots, S_n$ be prices and shares for the other categories, separability would imply that the underlying indirect utility function can be written as

$$\log V = F[\log V^1(P_2/M, P_3/M, \ldots, t), P_1/M, t]. \tag{2.4}$$

If the underlying indirect utility function is groupwise separable, then the following restrictions must hold:[13]

12. Note that stationarity is equivalent to explicit neutrality. An indirect utility function is explicitly neutral if it can be written as

$$\log V = \log V^1(P_1/M, P_2/M, \ldots, P_n/M) + F(t).$$

13. See Jorgenson and Lau (101) for a derivation of these restrictions.

$$\beta_{12} = \rho_1\alpha_2, \qquad \beta_{13} = \rho_1\alpha_3, \ldots, \beta_{1n} = \rho_1\alpha_n,$$

where $\rho_1$ is a constant. Even if the underlying indirect utility function is groupwise separable, the translog approximation need not be. Explicit groupwise separability ensures that the translog approximation is also groupwise separable. This requires the additional restriction that $\rho_1 = 0$.

We will also estimate share equations based on homothetic indirect utility functions. The indirect utility function is homothetic if it can be written as

$$\log V = F[\log H(P_1/M, \ldots, P_n/M), t], \tag{2.5}$$

where $H$ is homogeneous of degree 1. Under homotheticity the budget shares $S_j$ are independent of total expenditures $M$, which implies that the income elasticities of demand for every commodity are the same and equal to unity. There is no a priori basis for assuming homotheticity in our model of total consumption expenditures, but the model of fuel expenditures must be assumed to be homothetic to be consistent with a nonrecursive two-stage model of consumer spending. The underlying indirect utility function is homothetic if

$$\beta_{Mj} = \sigma\alpha_j, \qquad j = 1, \ldots, n,$$

where $\sigma$ is a constant. Explicit homotheticity will ensure that the translog approximation is also homothetic, and this requires the additional asumption that $\sigma = 0$. If the indirect utility function is explicitly homothetic and $\beta_{Mt} = 0$, then it is also homogeneous.

We will also test the restriction of explicit additivity. An indirect translog utility function is explicitly additive if it can be written in the form

$$\log V = \log V^1(P_1/M, t) + \ldots + \log V^n(P_n/M, t). \tag{2.6}$$

A necessary and sufficient condition for explicit additivity in the commodities is that the indirect translog function is explicitly groupwise separable in any pair of commodities from the remaining commodity. Therefore the parameter constraints implied by this restriction are that $\beta_{ij} = 0, i \neq j$.

We assume that the share equations have additive error terms that are independently and normally distributed. Under this assumption we can test various sets of parameter restrictions using a simple chi-square test. An appropriate asymptotic test

$$-2 \log \Lambda = N(\log |\hat{\Omega}_r| - \log |\hat{\Omega}_u|), \tag{2.7}$$

where $|\hat{\Omega}_r|$ and $|\hat{\Omega}_u|$ are the determinants of the estimated error covariance matrices for the restricted and unrestricted models respectively, and $N$ is the number of observations. In large samples this statistic is distributed as chi-square with degrees of freedom equal to the number of parameter restrictions being tested. We point out, however, that this chi-square test is a weak test for the acceptance of restrictions. If, for example, the test statistic for a particular set of parameter restrictions falls below some critical level, that only tells us that the data do not force us to reject the restrictions.

It is important to remember that there are only certain ranges of inputs over which the indirect translog utility function is a meaningful approximation to the underlying utility function. Consider, for example, the marginal utility of income (or of total expenditure), $\lambda = \partial V/\partial M$:

$$\lambda = \frac{V}{M}\frac{\partial \log V}{\partial \log M} = -\frac{V}{M}\sum_i \left(\alpha_i + \sum_j \beta_{ij} \log \frac{P_j}{M} + \beta_{ij} \cdot t\right). \tag{2.8}$$

The $\alpha_i$ sum to $-1$ by the normalization, while the $\beta_{ij}$ can be positive or negative. If some $\beta_{ij}$ are positive, then, as $M$ becomes zero, $\lambda$ can become negative, and if some $\beta_{ij}$ are negative, then, as $M$ becomes increasingly large, $\lambda$ can become negative. Thus there are ranges of input space for which the translog approximation may not be meaningful. It is important therefore to check estimated translog models by determining whether the marginal utility of income is positive over the range of historical (and forecasted) input data.

The rate at which the marginal utility of income declines as income increases is an interesting characteristic of the structure of demand. It is therefore useful to compute Frisch's welfare indicator for a model of consumption expenditures. This indicator is simply the income elasticity of the marginal utility of income, that is, $\eta_{\lambda M} = \partial \log \lambda / \partial \log M$. For a utility function that is well behaved over the entire input space (which the translog is not) $\eta_{\lambda M}$ would range from a large negative number (when $M$ is zero) to zero (as $M$

approaches infinity). Taking the log of equation (2.8) and differentiating, we have for the indirect translog function:[14]

$$\eta_{\lambda M} = -1 - \frac{\sum_i \sum_j \beta_{ij}}{\sum_i \alpha_i + \sum_i \sum_j \beta_{ij} \log \frac{P_j}{M} + \sum_i \beta_{it} \cdot t}. \tag{2.9}$$

As $M \to 0$, we see that $\eta_{\lambda M} \to -1$, so that a very small level of income is clearly out of the meaningful range. The same is true for $M \to \infty$. If the utility function is homothetic, however, this indicator is independent of income (but time-varying if the utility function is nonstationary). Since for most of the models that we will estimate the $\beta_{ij}$ and $\beta_{it}$ parameters will be held constant across countries, $\eta_{\lambda M}$ would also be constant across countries.

Income and price elasticities of demand can be determined from the estimated parameters of the share equations. The income elasticity of demand for good $j$, $\eta_{jM} = \partial \log X_j / \partial \log M$, is found by multiplying the share equation (2.3) by $M/P_j$ and differentiating

$$\eta_{jM} = 1 + \frac{\sum_i \beta_{Mi} - \sum_i \beta_{ji}/S_j}{\alpha_M + \sum_i \beta_{Mi} \log \frac{P_i}{M} + \beta_{Mt} \cdot t}. \tag{2.10}$$

Note that this is also the formula for the expenditure elasticity $\partial \log (P_j X_j) \partial \log M$. (Observe from equation (2.10) that if the indirect utility function is homothetic, $\eta_{jM} = 1$ for all $j$.) The own-price elasticity $\eta_{jj} = \partial \log X_j / \partial \log P_j$ is

$$\eta_{jj} = -1 + \frac{\beta_{jj}/S_j - \beta_{Mj}}{\alpha_M + \sum_i \beta_{Mi} \log \frac{P_i}{M} + \beta_{Mt} \cdot t}, \tag{2.11}$$

and the cross-price elasticities $\eta_{ji} = \partial \log X_j / \partial \log P_i$ are

$$\eta_{ji} = \frac{\beta_{ji}/S_j - \beta_{Mi}}{\alpha_M + \sum_i \beta_{Mi} \log \frac{P_i}{M} + \beta_{Mt} \cdot t}. \tag{2.12}$$

These elasticity formulas are based on the assumption that total expenditure stays constant. They are thus partial elasticities that

14. The same equation also applies to the direct translog utility function.

can be applied to each stage of our two-stage demand model, but they cannot be used to determine the total effect of a change in price or income on the demand for a particular fuel. If the price of oil changes, there will be a change in total expenditures on energy, and this will also affect the demand for oil. The total own-price elasticity $\eta^*_{jj} = d\log X_j/d\log P_j$ is given by

$$\eta^*_{jj} = \frac{P_j}{X_j}\left[\frac{\partial X_j}{\partial P_j} + \frac{\partial X_j}{\partial M_E}\frac{\partial M_E}{\partial P_E}\frac{\partial P_E}{\partial P_j}\right], \tag{2.13}$$

where $P_j$ and $X_j$ are the price and quantity of fuel $j$, $M_E$ is expenditures on energy, and $P_E$ is the price index for energy.

To determine the total elasticity we therefore need an expression for the price of energy in terms of the prices of individual fuels. Since fuels are not perfect substitutes, we cannot determine $P_E$ as a simple weighted average of the fuel prices. Instead we view $P_E$ as the cost of producing heat from fuel inputs, and use a translog cost function with constant returns to model this production process:

$$\log P_E = \gamma_0 + \sum_i \gamma_i \log P_i + \sum_i \sum_j \gamma_{ij} \log P_i \log P_j. \tag{2.14}$$

This is an energy price aggregator,[15] and can be determined up to a scalar $\gamma_0$ by estimating the share equations $S_i = \gamma_i + \Sigma_j \gamma_{ij} \log P_j$. Given equation (2.14) for the price of energy, we have

$$\frac{\partial P_E}{\partial P_j} = \frac{P_E}{P_j} S_j. \tag{2.15}$$

We can thus compute $\eta^*_{jj}$ from the fact that

$$\frac{\partial X_j}{\partial P_j} = \frac{X_j}{P_j} \eta_{jj}, \tag{2.16}$$

where $\eta_{jj}$ is the partial own-price elasticity for fuel $j$ given by equation (2.11),

$$\frac{\partial X_j}{\partial M_E} = \frac{X_j}{M_E} \eta_{jM_E}, \tag{2.17}$$

15. We will discuss the energy price aggregator in more detail in the next chapter.

where $\eta_{jM_E}$ is the expenditure elasticity for fuel $j$ given by equation (2.10), and

$$\frac{\partial M_E}{\partial P_E} = X_E(1 + \eta_{EE}), \tag{2.18}$$

where $X_E$ is the total quantity of energy consumed and $\eta_{EE}$ is the partial own-price elasticity of energy consumption. Now substituting (2.15), (2.16), (2.17), and (2.18) into (2.13), we have

$$\eta_{jj}^* = \eta_{jj} + \eta_{jM_E}(1 + \eta_{EE}). \tag{2.19}$$

We can similarly compute the total cross-price elasticity, $\eta_{ji}^*$, from

$$\eta_{ji}^* = \frac{P_i}{X_j}\left[\frac{\partial X_j}{\partial P_i} + \frac{\partial X_j}{\partial M_E}\frac{\partial M_E}{\partial P_E}\frac{\partial P_E}{\partial P_i}\right] = \eta_{ji} + S_i\eta_{jM_E}(1 + \eta_{EE}), \tag{2.20}$$

and the total income elasticity, $\eta_{jM}^*$, from

$$\eta_{jM}^* = \frac{M}{X_j}\frac{\partial X_j}{\partial M_E}\frac{\partial M_E}{\partial M} \tag{2.21}$$

Note that since $\partial M_E/\partial M = (M_E/M)\eta_{EM}$, where $\eta_{EM}$ is the income elasticity of energy expenditures, we obtain

$$\eta_{jM}^* = \eta_{jM_E} \cdot \eta_{EM}. \tag{2.22}$$

Since the indirect utility function for fuels is assumed to be homothetic, this reduces to $\eta_{jM}^* = \eta_{EM}$, that is, the total income elasticity of demand for each fuel is equal to the income elasticity of demand for energy.

At this point, let us summarize the steps involved in applying the indirect translog utility function to a two-stage model of residential energy demand. First, the share equations of (2.3) are estimated for expenditure shares corresponding to a breakdown of total consumption expenditures into energy and several other categories. (We will work with six categories of consumption expenditures: food, apparel, durable goods, transportation and communication, energy, and "all other.") These share equations can be estimated in their unrestricted form, so that even the parameter restrictions corresponding to utility maximization can be tested. Then, given the hypothesis of utility maximization, other restrictions can be tested by estimating the share equations with the particular parameter restrictions imposed and by comparing the results to the unrestricted estimate using the likelihood ratio test of equation (2.7).

Share equations are next estimated for the breakdown of energy expenditures into expenditures on oil, natural gas, coal, and electricity. Although homotheticity is a restriction that can be tested rather than assumed a priori for the breakdown of total consumption expenditures, it must hold for the breakdown of energy expenditures if consumers indeed make (simultaneous) separate decisions regarding the allocation of total consumption expenditures and the allocation of energy expenditures. For this reason we maintain the hypothesis of homotheticity in estimating share equations for fuel expenditures and test other restrictions (such as stationarity or additivity) conditional on homotheticity.

There are a number of methodological issues that must be resolved before this model of residential energy demand can be estimated, not the least of which is the choice of estimation method. Here, however, we concentrate on alternative theoretical specifications for our models, and we leave the discussion of these other issues to the next chapter.

**Dynamic Versions of the Indirect Translog Utility Function**

By estimating the static version of the translog model using pooled international time-series cross-section data, we can expect to obtain estimates of long-run energy demand elasticities. A problem with the static model, however, is that it cannot be used to explain differences between short-run and long-run elasticities. Even when the time trend is included in the indirect utility function, the model, although nonstationary, is not dynamic—tastes can change slowly over time, but there is no dynamic (lagged) response in demand to a sudden change in price. Thus adjustments in demand, while possibly nonstationary, are assumed to occur instantaneously.

Because of the limited scope of this study we have not empirically estimated demand models corresponding to a dynamic version of the indirect translog utility function. Nonetheless, here we discuss two basic approaches that can be used to introduce dynamic adjustments into the direct or indirect translog function.

The first approach involves specifying the translog approximation to the utility function (direct or indirect) to include lagged quantities, prices, or shares. The advantage of this approach is that adding up is always preserved in the resulting share equations without the introduction of additional parameter constraints. In addition, the share equations are still the result of an optimization process—although the optimization is static despite the introduction of lagged variables. The disadvantage is that the translog approximation makes the dynamic specification somewhat arbitrary. (There is, for example, no explicit linkage to a turnover in the appliance stock.)

As an example of this approach, we could write the indirect translog utility function as

$$\log V = \alpha_0 + \sum_i \alpha_i \log (P_i/M) + \frac{1}{2} \sum_i \sum_j \beta_{ij} \log (P_i/M) \log (P_j/M) + \sum_i d_i \log (P_i/M) D_{i,t-1}, \quad (2.23)$$

where $D_{i,t-1}$ is a lagged term in price, quantity, or share that is considered an exogenous input to the determination of current share.[16] Logical choices for $D_{i,t-1}$ would be the quantity $X_{i,t-1}$ or the share $S_{i,t-1}$.[17] Applying Roy's identity as given by equation (2.1) to equation (2.23) yields the share equations:

$$S_j = \frac{P_j X_j}{M} = \frac{\alpha_j + d_j D_{j,t-1} + \sum_i \beta_{ji} \log (P_i/M)}{\alpha_M + \sum_i d_i D_{i,t-1} + \sum_i \beta_{Mi} \log (P_i/M)}, \quad j = 1, \ldots, n, \quad (2.24)$$

where $\alpha_M$ and $\beta_{Mi}$ are defined as before. Note that unless all of the $d_i$ are zero, the homothetic form of equation (2.24)—for which the $\beta_{Mi}$ are zero—is nonlinear in the parameters. As a result estimation of (2.24) can be computationally costly, even under the assumption of homotheticity. The shares in (2.24) will always add to 1, however,

16. We are assuming that consumers determine their budget shares via static utility maximization, that is, they maximize utility at each instant of time ignoring the future rather than maximizing the sum over time of discounted utilities. The $D_{i,t-1}$ (together with current prices and income) simply represent the current state of the world. As shown by Hoel (85), even in a static model dynamic utility maximization can result in a different marginal utility of income.

17. The form of equation (2.23) using lagged quantity $X_{i,t-1}$ was suggested by Manser (115, 116), who applied it to the estimation of food demand.

even if lagged shares are used as the $D_{i,t-1}$, and, assuming that the errors are not serially correlated, the parameter estimates will be invariant to the choice of share that is dropped.

A second approach that can be used is to introduce the dynamic adjustment directly into the share equations. This has the advantage of facilitating the use of simple and intuitively pleasing adjustment mechanisms. It has the disadvantage that adding up will not be preserved unless additional (and possibly highly restrictive) parameter restrictions are introduced.

We consider two examples of this approach. In the first, each quantity is assumed to adjust to a desired level:

$$X_{i,t} = X_{i,t-1} + \delta_i(X^*_{i,t} - X_{i,t-1}), \tag{2.25}$$

where $X^*_{i,t}$ is the desired quantity of commodity $i$ as determined from static utility maximization, and $\delta_i$ is an adjustment parameter. This yields the share equations

$$S_{j,t} = \delta_j S^*_{j,t} + (1 - \delta_j)S_{j,t-1}\left[\frac{P_{j,t}/M_t}{P_{j,t-1}/M_{t-1}}\right], \quad j = 1, \ldots, n, \tag{2.26}$$

or, using the indirect translog function for $S^*_{j,t}$,

$$\frac{\alpha_j + \sum_i \beta_{ji} \log (P_i/M)}{\alpha_M + \sum_i \beta_{Mi} \log (P_i/M)} + (1 - \delta_j)S_{j,t-1}\left[\frac{P_{j,t}/M_t}{P_{j,t-1}/M_{t-1}}\right]. \tag{2.27}$$

The parameters of the share equations (2.27) are estimated subject to the constraints $\beta'_{ij} = \beta'_{ji}$, $\beta_{Mj} = \Sigma_k \beta'_{kj}/\delta_k$ are the same in each equation, and $\alpha_M = \Sigma_k \alpha'_k/\delta_k = -1$.

Note that the shares $S_{j,t}$ need not add to 1. Adding up can be imposed by estimating only $n - 1$ of the share equations, and determining the parameters of the $n$th equation from $\Sigma S_{j,t} = 1$, but the estimated parameters will depend on the particular equation that is not estimated. Despite this deficiency, however, the specification of equation (2.27) permits the introduction of dynamic adjustments in a simple and appealing manner.

Alternatively, we can assume that the shares adjust to the desired shares as follows:[18]

18. This approach is discussed in some detail in Berndt, Fuss, and Waverman (18).

$$S_{i,t} = S_{i,t-1} + \sum_j \delta_{ij}(S^*_{j,t} - S_{j,t-1}). \tag{2.28}$$

Adding up requires that the sum of all changes in shares be zero, that is,

$$\sum_i (S_{i,t} - S_{i,t-1}) = 0,$$

so that

$$\sum_i \sum_j \delta_{ij}(S^*_{jt} - S_{j,t-1}) = \sum_j (S^*_{j,t} \sum_i \delta_{ij} - S_{j,t-1} \sum_i \delta_{ij}) = 0. \tag{2.29}$$

Since the $S^*_{j,t}$ and $S_{j,t-1}$ sum to 1, this equation implies the necessary condition that *all of the columns of the matrix* $(\delta_{ij})$ *sum to the same arbitrary constant,* in other words,

$$\mathbf{l}'\boldsymbol{\delta} = c\mathbf{l}' \tag{2.30}$$

where $\mathbf{l}$ is a vector of 1's (ones), $\boldsymbol{\delta}$ is the matrix $(\delta_{ij})$, and $c$ is an arbitrary constant.[19] Note that if the number of shares is greater than 2, there are alternative constraints on the $\delta_{ij}$ that can be imposed to satisfy (2.30). In addition, equation (2.30) implies that $\boldsymbol{\delta}$ cannot be diagonal unless the adjustment coefficients for every share are the same, so that the adjustment of the $i$th commodity share would generally not depend only on that share but on other shares as well.

**Multinomial Logit Models for Fuel Choice**

Multinomial logit models have already been used to study the breakdown of energy consumption into demands for individual fuels in the U.S.[20] and Canada.[21] Although the logit model is not based on assumptions of utility maximization, it has properties that make it appealing as a basis for describing interfuel competition. The model is consistent in terms of shares adding to 1, and shares respond to price changes in a way that is intuitively appealing; as the share of,

19. For a discussion of adding up conditions for more general lag structures, see Berndt and Savin (20), and Berndt, Fuss, and Waverman (18).
20. See Baughman and Joskow (11) and Joskow and Baughman (102). For the specification of a multinomial logit model of fuel choice based on changes in appliance stocks, see Pindyck (133).
21. See Fuss and Waverman (62).

say, natural gas becomes small, it requires increasingly large price changes to make it still smaller. Also, the logit model is relatively easy to estimate; as long as the share data represents aggregated samples of individuals' decisions (that is, average shares for a large number of consumers) rather than individual decisions, ordinary or generalized least squares can be used. Another advantage of the logit model is that it allows considerable flexibility for working a dynamic structure into the specification.

The logit model also has disadvantages, however. Estimates become inefficient when there are zeros in the share data. In addition, all cross elasticities for a given price are equal; as Hausman and Wood (81) have shown, they are the sum of the price elasticity for total expenditure minus the own-price elasticity weighted by the share. This fact that cross elasticities are determined by total and own elasticities is restrictive, but less so than the restrictions inherent in the linear expenditure system, additive quadratic model, and other consistency models.

Here we specify static and dynamic versions of the multinomial logit model as descriptions of interfuel substitution. Despite its limitations, the static model provides a check on the plausibility of our estimates from the translog model, and the dynamic model provides a means of testing whether the data contain enough timewise variation to permit us to elicit estimates of the adjustment times inherent in interfuel substitution. (Unfortunately, our estimation of the model did not succeed in eliciting this information. Estimation results are presented in the appendix to chapter 4.)

We can write the logit model for the four fuel breakdown—oil, gas, coal, and electricity—as follows:

$$\frac{Q_i}{Q_T} = \frac{\exp f_i(\mathbf{x}\boldsymbol{\beta})}{\sum_{j=1}^{4} exp\ f_j(\mathbf{x}\boldsymbol{\beta})}, \tag{2.31}$$

where $Q_i$ is the quantity in thermal units such as Tcals of fuel $i$, $Q_T = \Sigma Q_i$, and the $f_i$ are functions of a vector of attributes $\mathbf{x}$ and vector of parameters $\boldsymbol{\beta}$.[22] Given this model, the relative shares of any two fuels can be represented as

22. In effect we are assuming that consumer preferences are represented by a choice index which for the $i$th fuel, has the form $f_i(\mathbf{x}\boldsymbol{\beta}) + \epsilon_i(\mathbf{x})$, where

$$\log (Q_i/Q_j) = \log (S_i/S_j) = f_i(\mathbf{x}\boldsymbol{\beta}) - f_j(\mathbf{x}\boldsymbol{\beta}). \tag{2.32}$$

Note that only three equations are estimated since the parameters of the fourth equation are determined from the adding up constraint.

In estimating fuel shares, we include as attributes the relative price of each fuel. The relative oil price, for example, is the ratio of the real price of oil to the real price of energy, the latter being measured by a translog energy price aggregator, as discussed earlier. Other attributes may include per capita income, average temperature, and lagged quantity variables that allow shares to adjust dynamically to changes in price. Functional forms for the $f_i$ are somewhat arbitrary, but in the simplest model they might be linear functions of the relative fuel prices $\tilde{P}_i = P_i/P_E$, where $P_E$ is the aggregate price of energy, as well as income $Y$ and temperature $T$:

$$f_i(\mathbf{x}\boldsymbol{\beta}) = a_i + b_i\tilde{P}_i + c_iY + d_iT. \tag{2.33}$$

This yields the three estimating equations

$$\log (S_i/S_4) = (a_i - a_4) + b_i\tilde{P}_i - b_4\tilde{P}_4 + (c_i - c_4)Y + (d_i - d_4)T, \quad i = 1, 2, 3. \tag{2.34}$$

Note that these three equations must be estimated simultaneously, with $b_4$ constrained to be the same in each equation.

We can modify the logit model so that it represents preference functions that are dynamic (for example, through the effects of habit formation or stock adjustments). This can be done by including the lagged shares in the functions $f_i$:

---

$\epsilon_i$ is an error term. Then the probability that a consumer would choose fuel $i$ is

$$P_i = \text{Prob}[f_i(\mathbf{x}\boldsymbol{\beta}) + \epsilon_i(\mathbf{x}) > f_j(\mathbf{x}\boldsymbol{\beta}) + \epsilon_j(\mathbf{x})] \quad \text{for} \quad i \neq j.$$

If the error terms $\epsilon_i(\mathbf{x})$ are independently and identically distributed with the Weibull distribution

$$\text{Prob}[\epsilon_i(\mathbf{x}) < \epsilon] = e^{-e^{-\epsilon}},$$

then the probability that fuel $i$ will be chosen is given by equation (2.31). For further discussion, see McFadden (117), Domencich and McFadden (56), Cox (47), Theil (157), and chapter 8 of Pindyck and Rubinfeld (140). For an interesting application to aggregate demand analysis, see Park (127).

$$f_i(\mathbf{x}\boldsymbol{\beta}) = a_i + b_i\tilde{P}_i + c_iS_{i,t-1}. \qquad (2.35)$$

The three estimating equations are then

$$\log (S_i/S_4) = (a_i - a_4) + b_i\tilde{P}_i - b_4\tilde{P}_4 + c_iS_{i,t-1} - c_4S_{4,t-1},$$
$$i = 1, 2, 3. \qquad (2.36)$$

Note that two lagged shares appear in each equation. The three equations must again be estimated simultaneously, with both $b_4$ and $c_4$ constrained to be the same across the equations.[23]

## 2.2 The Industrial Sector

The effects of GNP growth and changing fuel prices on the industrial demand for energy depend on the substitutability of energy and other factors of production and the substitutability of fuels within the energy aggregate. Until recently, most studies of production concentrated on the substitutability of capital and labor, assuming that in the underlying production function these factors were separable from energy and raw material inputs. Dramatic increases in

23. It is straightforward to calculate income and price elasticities of shares for the static logit model. From equation (2.34) we can obtain the income elasticity for a model with $n$ shares as follows:

$$dS_k/S_k - dS_n/S_n = (c_k - c_n)dY.$$

Since $\Sigma\, dS_k = 0$, we have that

$$\sum_{k=1}^{n-1} dS_k = \sum_{k=1}^{n-1}\left\{\frac{S_k}{S_n}\, dS_n + S_k(c_k - c_n)dY\right\} = -\, dS_n,$$

or, $$dS_n/S_n = c_ndY(1 - S_n) - dY\sum_{k=1}^{n-1} c_kS_k.$$

After some manipulation, this reduces to

$$\eta^S_{iY} = \frac{dS_i/S_i}{dY/Y} = (c_i - c_n)Y - \sum_{k=1}^{n-1}(c_k - c_n)S_kY.$$

To obtain the own-price elasticities of shares, note that

$$dS_k/S_k - dS_n/S_n = b_kdP_k.$$

Then using the fact that $\Sigma\, dS_k = 0$ and $\Sigma\, S_k = 1$, we can obtain

$$\eta^S_{ii} = \frac{dS_i/S_i}{dP_i/P_i} = b_i(1 - S_i)P_i, \qquad i = 1, \ldots, n.$$

the price of energy, together with the appearance of studies such as that of Berndt and Wood (22) which indicated that energy and capital may in fact be complements rather than substitutes, resulted in an increased interest in the substitutability of energy with other factors of production. In addition, rapid changes in the prices of individual fuels raised the issue of how rapidly and to what extent the different fuels used in the industrial sector could be substituted for each other. Finally, there has been an increased concern with the growth in energy demand itself, brought about by growth in industrial activity.

Several recent studies have examined the role of energy in the structure of production, but the evidence on factor and fuel substitutability is mixed. Berndt and Wood (22), Hudson and Jorgenson (94), Fuss (60), and Magnus (114) all worked with time-series data for a single country, and found energy and labor to be substitutes, but energy and capital to be complements. Griffin and Gregory (71), using cross-sectional data at five-year intervals for nine countries to capture long-run effects, found energy and capital to be substitutes. Fuss (60) found for Canada moderate substitutability between coal, gas, and oil, but almost none between these fuels and electricity, while Halvorsen (77) found greater substitutability among all fuels in the U.S.[24]

Here we will take another look at the role of energy in the structure of production. Our objectives are to provide some new evidence on the extent of capital, labor, and energy substitutability, the long-run own- and cross-price elasticities of the demand for energy and the demands for individual fuels, the impact of growth in industrial activity on the demands for energy and individual fuels, and for different countries the effects of increased energy costs on the cost of output.

As in the residential sector, we treat industrial energy demand using a two-stage approach. In the first stage energy demand is a derived factor demand (assuming that factor inputs are chosen to minimize the total cost of production) based on a translog cost function. Factor inputs include capital, labor, and energy; a lack of data makes it impossible to include materials as an explicit input,

24. We survey the results of these and other studies of industrial energy demand at the end of chapter 5.

so we must assume separability of materials from the other factors in the underlying structure of production. In the second stage expenditures on energy are broken down into expenditures on oil, natural gas, coal, and electricity, now under the assumption that fuel inputs are chosen to minimize the cost of energy. Here translog cost functions are again used, but we have also experimented with the use of the multinominal logit model to explain fuel shares. The use of this two-stage approach requires certain additional assumptions about the underlying structure of production. In particular, we must assume that the production function is homothetically separable in the capital, labor, and energy aggregates—in other words, first, that expenditure shares for fuels are independent of the expenditure shares for capital and labor, and second, that expenditure shares for fuels are independent of total energy expenditures.[25]

As was the case with the indirect translog utility function in the residential sector, the use of the translog cost function for industrial demand has the advantage that it permits us to obtain relatively unrestricted estimates of elasticities of substitution and demand elasticities. We need not a priori assume that the cost function is homothetic (at least in modelling the demand for capital, labor, and energy inputs), and we need not assume that the elasticities of substitution between different inputs are all the same. Of course, there are other generalized cost and production functions which could be used that also impose no a priori restrictions on elasticities of substitution, and one or more of these alternative functions might provide tighter estimates than does the translog function.[26] Here, however, we work only with the translog function.

25. The necessity for these assumptions is shown by Berndt and Christensen (16) and Denny and Fuss (51). Note that this second assumption of homotheticity of fuel shares will not be violated in the use of a logit model if total energy expenditures is not an explanatory variable in that model.
26. For example, the generalized Leontief function and the generalized Cobb-Douglas function, both of which were introduced by Diewert (52,53). Lau and Tamura (110) used the generalized Leontief production function in estimating input substitution in petrochemical refining. Magnus (114) used the generalized Cobb-Douglas cost function in estimating substitution between capital, labor, and energy inputs in Dutch manufacturing.

Our models are estimated using pooled time-series cross-section data for ten countries: Canada, France, Italy, Japan, the Netherlands, Norway, Sweden, the U.K., the U.S., and West Germany. As in the residential model, pooling countries is essential, because our objective is to identify long-run elasticities. We also estimate the models using alternative pooling in order to identify possible cross-country differences in elasticities.

We begin here by reviewing the properties of the static translog cost function, and discuss its application to the industrial demand model. Although we do not estimate them, we next describe some alternative dynamic specifications of the translog cost function; these specifications permit nonconstant returns to scale in the short run, with an adjustment to constant returns in the long run. Finally, we briefly discuss the use of the multinominal logit model as an alternative means of describing fuel shares.

**The Structure of Production**

Our approach requires certain assumptions about the structure of production. First, we assume that capital, labor, and energy inputs are as a group weakly separable from the fourth input, materials. Materials includes intermediate inputs as well as nonenergy raw materials. Weak separability here means that the marginal rate of substitution between any two of the first three inputs is independent of the quantity of materials used as an input. This is a necessary and sufficient condition for the production function to be of the form $Q = F[f(K, K, E); M]$.[27] This assumption is made necessary by the fact that we have no data from which to construct price indices of materials inputs, and therefore we can only estimate unrestricted elasticities of substitution between capital, labor, and energy.

Second, we assume that the production function is weakly separable in the major categories of capital, labor, and energy.[28] This implies that the marginal rates of substitution between individual

27. For a proof and further discussion, see Berndt and Christensen (16).
28. Halvorsen and Ford (79) recently used translog cost functions to test for separability of the energy aggregate for each of eight individual two-digit industries in the U.S. They found separability to hold for four of the eight industries.

fuels is independent of the quantities of capital and labor. The assumption permits us to use aggregate price indices for capital, labor, and energy inputs—in particular to construct an energy price index that aggregates the price of the four fuels and a price index of capital services that aggregates different types of capital.

Finally, we assume that the capital, labor, and energy aggregates are homothetic in their components—in particular that the energy aggregate is homothetic in its oil, gas, coal, and electricity inputs.[29] As shown by Denny and Fuss (51), this last assumption provides a necessary and sufficient condition for an underlying two-stage optimization process, that is, optimizing the mix of fuels that make up the energy input, and then optimally choosing quantities of capital, labor, and energy. Equivalently we can express these three assumptions by writing the production function as

$$Q = F[(K, L, e(F_1, F_2, F_3, F_4)); M], \tag{2.37}$$

where $e$ is a homothetic function of the four fuels.

As shown by Shepard (149), if the factor prices and output level are exogenously determined, the production structure described by (2.37) can alternatively be described by a cost function that is also weakly separable, that is, a function of the form

$$C = G[g(P_K, P_L, P_E(P_{F1}, P_{F2}, P_{F3}, P_{F4})Q); P_M, Q]. \tag{2.38}$$

Here $P_E$ is an aggregate price index of energy, that is, a function that aggregates the fuel prices $P_{Fi}$. This aggregator function is homothetic, and thus does not include the total quantity of energy as one of its arguments.

It is important to recognize that this cost function need not be the same for all industries. As Halvorsen and Ford (79) recently found in examining cost functions for capital, labor, energy, and materials inputs, and as Moroney and Toevs (119) found in examining cost functions for capital, labor, and all raw materials (including energy) inputs, there is considerable evidence that the cost function, and the implied elasticities of factor substitution, vary considerably across industries. The use of interindustry data, however, would be impractical for an international study such as this

29. The second and third assumptions together are referred to as homothetic separability.

one. Nonetheless, we must be aware of the fact that we may be introducing a source of bias by aggregating over all industries.

**Use of the Translog Cost Function**

Our approach here is similar to that used recently by Fuss and Waverman (62) and by Fuss (60) in estimating the demand for energy in Canadian manufacturing. We first represent the price of energy (which is the unit cost of energy to a producer choosing fuel inputs) by a homothetic translog cost function with constant returns to scale. Estimation of the share equations implied by this cost function gives us the own and cross partial price elasticities for the four fuels. In addition, the cost function itself provides an instrumental variable for the price of energy. The second step is to represent the cost of industrial output by a nonhomothetic translog cost function. Estimation of the share equations implied by this cost function gives us the elasticities of substitution and the own- and cross-price elasticities for capital, labor, and energy.

It is important to point out that we could have chosen to use translog production functions rather than cost functions in estimating elasticities. Since the translog cost and production functions are not self-dual, different elasticity estimates would result, and as Burgess (26) has recently shown, the difference could be significant. However, we choose to use the cost function because it is more appropriate to take prices as exogenous than quantities.

We begin by reviewing the properties of the translog cost function.[30] The translog cost function is a second-order approximation to an arbitrary cost function, and has the form

$$\log C = \alpha_0 + \alpha_Q \log Q + \sum_i \alpha_i \log P_i + \frac{1}{2}\gamma_{QQ}(\log Q)^2 + \frac{1}{2} \sum_i \sum_j \gamma_{ij} \log P_i \log P_j + \sum_i \gamma_{Qi} \log Q \log P_i, \qquad (2.39)$$

30. The translog production function and cost function were introduced by Christensen, Jorgenson, and Lau (40, 41). Applications of the translog cost function can be found in the work of Berndt and Christensen (15), Berndt and Wood (22), Christensen and Greene (38), Fuss and Waverman (62), Fuss (60), Griffen and Gregory (71), Halvorsen (77), and Hudson and Jorgenson (94). The translog production function has been used by Humphrey and Moroney (96) and Moroney and Toevs (119, 120).

where $C$ is total cost, $Q$ is output, and $P_i$ are the factor prices. From Shepard's lemma (149), the derived demand functions are found by differentiating the cost function with respect to the prices, that is, $X_i = \partial C/\partial P_i$. Thus the share equations are given by $S_i = \partial \log C/\partial \log P_i = (P_iX_i)/C$, or

$$S_i = \alpha_i + \gamma_{Qi} \log Q + \sum_j \gamma_{ij} \log P_j, \qquad i = 1, \ldots, n. \tag{2.40}$$

Since the shares must add to 1, only $n - 1$ of the share equations need be estimated. Note, however, that the parameters $\alpha_0$, $\alpha_Q$, and $\gamma_{QQ}$ are not identified unless the cost function itself is estimated.[31]

The consistent estimation of the share equations (2.40) requires the exogeneity of both the prices $P_j$ and output $Q$. It is indeed reasonable to assume that factor prices are exogenous, and, as mentioned above, this is one of the reasons we work with a translog cost function rather than the production function. We should keep in mind, however, that it may be less valid to assume that total output $Q$ is exogenous.

The cost function must be homogeneous of degree 1 in prices, in other words, a proportional increase in all prices must result in the same proportional increase in the total cost of output. Also, the cost function must satisfy the conditions corresponding to a well-behaved production function. Christensen, Jorgenson, and Lau (41) show that this implies that the following parameter restrictions must be imposed:

$$\sum_i \alpha_i = 1$$

$$\sum_i \gamma_{Qi} = 0$$

$$\gamma_{ij} = \gamma_{ji}, \qquad i \neq j$$

$$\sum_i \gamma_{ij} = \sum_j \gamma_{ij} = 0.$$

31. Corbo and Meller (46) estimated the translog production function directly (instead of the derived share equations) using capital and labor input data of individual firms. This allowed them to test whether the underlying production function is really translog and to test for competitive behavior.

Note that the cost function specified thus far is nonhomothetic, and may have nonconstant returns to scale. The cost function would be homothetic if it could be written as a separable function of output and factor prices. Thus the following parameter restrictions can be added to impose homotheticity:

$\gamma_{Qi} = 0.$

The cost function is also homogeneous if the elasticity of cost with respect to output ($\partial \log C/\partial \log Q$) is constant. This implies the additional restriction:

$\gamma_{QQ} = 0.$

Finally, we could impose the restriction that the elasticities of substitution between all factors are equal to 1 (so that the cost function corresponds to a Cobb-Douglas production function). This implies the additional parameter restrictions:

$\gamma_{ij} = 0.$

Rather than impose these restrictions a priori, we can test them using the simple chi-square test discussed earlier. Recall that under the assumption that the share equations have additive error terms that are normally and independently distributed, an appropriate asymptotic test statistic is given by equation (2.7), and is distributed as chi-square with degrees of freedom equal to the number of parameter restrictions being tested.

A convenient way of describing the substitutability of various factors of production is through the "Allen partial elasticity of substitution," (see Allen, 4). This elasticity measures the percent change in the ratio of two factors resulting from a 1 percent change in their relative prices. Thus an elasticity of substitution of 1 for capital and labor implies that a 1 percent increase in the price ratio $P_L/P_K$ results in a 1 percent increase in the factor ratio $K/L$. If the Allen elasticity of substitution between two factors is positive, we call the factors substitutes (or AES substitutes), while if the elasticity is negative, the factors are complements.

It is important to keep clear the meaning of AES—it is a measure of relative substitution between two factors compared with the substitution effects of other factors. Take three factors, capital, labor, and energy, as an example. As Hogan (86) and Berndt and

Wood (24) point out, with total output fixed, two inputs (say capital and energy) taken by themselves are always substitutes. However, an increase in the price of energy, while leading to the substitution of capital, might result (with a decrease in output) in greater substitution of labor, a decrease in the use of capital and energy together, and a net decrease in the use of capital.[32] Thus, if the substitution between capital and energy is small relative to the substitution of their composite with labor, capital and energy are AES complements (or, in the Berndt and Wood terminology, "net complements"). If, on the other hand, the substitution between capital and energy is large relative to the substitution of their composite with labor, then capital and labor are AES substitutes (or net substitutes).

Uzawa (162) showed that the Allen elasticities of substitution $\sigma_{ij}$ can be computed from the formula

$$\sigma_{ij} = \frac{C(\partial^2 C/\partial P_i \partial P_j)}{(\partial C/\partial P_i)(\partial C/\partial P_j)}$$

By applying this to the translog cost function of equation (2.39), we obtain the following expressions giving the elasticities of substitution in terms of the estimated parameters and estimated expenditure shares:

$$\begin{aligned} \sigma_{ij} &= (\gamma_{ij} + S_i S_j)/S_i S_j, \qquad i \neq j, \\ \sigma_{ii} &= [\gamma_{ii} + S_i(S_i - 1)]/S_i^2. \end{aligned} \tag{2.41}$$

32. Hogan (86) uses a nice example to make this clear: suppose the production function is of the form

$$F(K, L, E) = G[H(K, E), L],$$

with both $G$ and $H$ constant elasticity of substitution production functions. Then $K$ and $E$ are substitutes within $H$, and $H$ and $L$ are substitutes within $G$. However, $K$ and $E$ might still be AES complements in $G$. The AES between capital and energy is given by

$$\sigma_{KE} = \sigma_G + (\sigma_H - \sigma_G)/S_H,$$

where $\sigma_H$ and $\sigma_G$ are the elasticities of substitution in the functions $H$ and $G$, and $S_H$ is the expenditure share of $H$ in $G$. Thus if $\sigma_G > \sigma_H/(1 - S_H)$, capital and energy are AES complements. The concepts of complementarity and substitutability when there are more than two factors of production are also discussed by Sato and Koizumi (148).

It is easy to show that the own- and cross-price elasticities of demand are related to the elasticities of substitution as follows:

$$\eta_{ii} = \partial \log X_i / \partial \log P_i = \sigma_{ii} S_i,$$
$$\eta_{ij} = \partial \log X_i / \partial \log P_j = \sigma_{ij} S_j. \tag{2.42}$$

These are partial price elasticities; when applied to fuels they account only for substitution between fuels, under the constraint that the total quantity of energy consumed remains constant. (Note, however, that expenditures on energy will not remain constant.) In fact, if the price of a particular fuel increases, the demand for that fuel will decrease for two reasons: interfuel substitution, resulting from changing relative fuel prices, and a decreased use of all energy, resulting from an increase in the aggregate price of energy. The total own-price elasticity for each fuel $\eta_{ii}^* = d\log X_i / d\log P_i$ accounts for both interfuel substitution and the effect of a change in the price of a fuel on total energy consumption, and is given by

$$\eta_{ii}^* = \frac{P_i}{X_i} \left[ \frac{\partial X_i}{\partial P_i} \bigg|_{E \text{ const.}} + \frac{\partial X_i}{\partial E} \frac{\partial E}{\partial P_E} \frac{\partial P_E}{\partial P_i} \right], \tag{2.43}$$

where $E$ is the total quantity of energy consumed and $P_E$ is the price index for energy.

Total fuel-price elasticities can thus be determined given an expression for the aggregate price of energy. The price of energy is given by the homothetic translog cost function with constant returns to scale:

$$\log P_E = \alpha_0 + \sum_i \alpha_i \log P_i + \sum_i \sum_j \gamma_{ij} \log P_i \log P_j. \tag{2.44}$$

This implies the fuel-share equations $S_i = \gamma_i + \Sigma \gamma_{ij} \log P_j$, so that we have

$$\frac{\partial P_E}{\partial P_i} = \frac{P_E}{P_i} S_i, \tag{2.45}$$

$$\frac{\partial E}{\partial P_E} = \frac{E}{P_E} \eta_{EE}, \tag{2.46}$$

where $\eta_{EE}$ is the own-price elasticity of energy, and, since the energy cost function is homothetic,

$$\frac{\partial X_i}{\partial E} = \frac{\partial X_i}{\partial M_E} \frac{M_E}{\partial E} = \frac{X_i}{M_E} P_E = \frac{X_i}{E}, \tag{2.47}$$

where $M_E = P_E \cdot E$ is total expenditures on energy. Then by substituting (2.45), (2.46), and (2.47) into (2.43), we obtain

$$\eta_{ii}^* = \eta_{ii} + \eta_{EE} S_i . \tag{2.48}$$

Similarly, we can compute the total cross-price elasticity $\eta_{ij}^*$ from

$$\eta_{ij}^* = \frac{P_j}{X_i} \left[ \left. \frac{\partial X_i}{\partial P_j} \right|_{E \text{ const.}} + \frac{\partial X_i}{\partial E} \frac{\partial E}{\partial P_E} \frac{\partial P_E}{\partial P_j} \right] = \eta_{ij} + \eta_{EE} S_j , \tag{2.49}$$

and the total output elasticity from

$$\eta_{iQ}^* = \frac{d\log X_i}{d\log Q} = \frac{Q}{X_i} \frac{\partial X_i}{\partial M_E} \frac{\partial M_E}{\partial Q} . \tag{2.50}$$

Here we are assuming that the value of output is equal to the value (cost) of inputs. This would be the case under perfect competition, or under oligopoly pricing based on a fixed percentage markup over cost. Otherwise this elasticity would be better referred to as a total cost elasticity, that is, the percent change in the demand for fuel $i$ corresponding to a 1 percent change in the total cost of production. Since the energy cost function is homothetic, equation (2.50) reduces to

$$\eta_{iQ}^* = \eta_{EQ} , \tag{2.51}$$

where $\eta_{EQ}$ is the elasticity of energy with respect to output changes. This elasticity can in turn be computed as follows:

$$\begin{aligned} \eta_{EQ} &= \frac{d\log E}{d\log Q} = \frac{d\log E}{d\log C} \cdot \frac{\partial \log C}{\partial \log Q} \\ &= \left[ \frac{\partial \log S_E}{\partial \log Q} \frac{\partial \log Q}{\partial \log C} + 1 \right] \frac{\partial \log C}{\partial \log Q} \\ &= \frac{\partial \log S_E}{\partial \log Q} + \frac{\partial \log C}{\partial \log Q} . \end{aligned} \tag{2.52}$$

Obtaining these derivatives from equations (2.39) and (2.40), we have

$$\eta_{EQ} = \frac{\gamma_{QE}}{S_E} + \alpha_Q + \gamma_{QQ} \log Q + \sum_{i=K}^{L,E} \gamma_{Qi} \log P_i . \tag{2.53}$$

Since the cost function is not estimated directly, we assume that $\alpha_Q$ and $\gamma_{QQ}$ are 1 and 0 respectively.

The total cost function need not exhibit constant returns to scale (although constant returns to scale is not inconsistent with a non-homothetic cost function), so it is useful to compute an index that measures the extent of scale economies. We follow Christensen and Greene (38) in defining the following index of scale economies (SCE):

$$\text{SCE} = 1 - \frac{\partial \log C}{\partial \log Q} = 1 - (\alpha_Q + \gamma_{QQ} \log Q + \sum_i \gamma_{Qi} \log P_i). \tag{2.54}$$

Note that if SCE is positive (negative), there is increasing (decreasing) returns to scale. SCE is a useful index and has a natural interpretation in percentage terms. However, it can only be computed if $\alpha_Q$ and $\gamma_{QQ}$ are known (which means estimating the cost function) or are assumed to be 1 and 0, respectively.

Finally, it is useful to calculate the elasticity of the average cost of production with respect to the price of energy, that is, $\eta_{CE} = \partial \log (AC)/\partial \log P_E$, and the elasticities of the average cost of production with respect to the prices of each fuel, that is, $\eta_{Ci} = \partial \log (AC)/\partial \log P_i$. This will enable us to calculate the effect of a 1-percent change in the price of energy, or a 1-percent change in the price of a single fuel, on the cost of industrial output. We follow Fuss (60) in calculating point elasticities for $\eta_{CE}$ and $\eta_{Ci}$. From equation (2.39) we have, assuming the prices of capital and labor are fixed,

$$\eta_{CE} = \alpha_E + \gamma_{EE} \log P_E + \gamma_{EK} \log P_K + \gamma_{EL} \log P_L + \gamma_{QE} \log Q. \tag{2.55}$$

We obtain $\eta_{Ci}$ from

$$\eta_{Ci} = \frac{\partial \log (AC)}{\partial \log P_E} \frac{\partial \log P_E}{\partial \log P_i} = \eta_{CE} S_i. \tag{2.56}$$

All of the elasticities can be evaluated numerically given sets of parameter estimates for the total cost function and energy cost function. Elasticity estimates are more useful, however, if they are accompanied by standard errors. A problem in calculating standard errors for our estimates of elasticities is that the elasticities are nonlinear functions of the estimated parameters (since the shares are themselves functions of the parameters). Thus there is no straightforward way to calculate elasticities exactly without reverting to Monte Carlo simulation. However, we can obtain approximate estimates of the standard errors under the assumption that the shares $S_i$ are constant and equal to the means (over the estimation time bounds) of their estimated values. Under this assumption we have, asymptotically,

$$\begin{aligned}
\text{Var}(\hat{\sigma}_{ij}) &= \text{Var}(\hat{\gamma}_{ij})/\hat{S}_i^2\hat{S}_j^2 \\
\text{Var}(\hat{\sigma}_{ii}) &= \text{Var}(\hat{\gamma}_{ii})/\hat{S}_i^4 \\
\text{Var}(\hat{\eta}_{ij}) &= \text{Var}(\hat{\gamma}_{ij})/\hat{S}_i^2 \\
\text{Var}(\hat{\eta}_{ii}) &= \text{Var}(\hat{\gamma}_{ii})/\hat{S}_i^2.
\end{aligned} \tag{2.57}$$

Let us now review the steps involved in applying the translog cost function to a model of industrial energy demand. First, the fuel-share equations

$$S_i = \alpha_i + \sum_j \gamma_{ij} \log P_j$$

are estimated, and the estimated parameters are used to calculate partial price elasticities for the four fuels. These equations are estimated subject to the parameter restrictions $\Sigma_i\, \alpha_i = 1$, $\gamma_{ij} = \gamma_{ji}$, and $\Sigma_i\, \gamma_{ij} = \Sigma_j\, \gamma_{ij} = 0$. We also test the additional restrictions $\gamma_{ij} = 0$. Then the estimated values of the $\alpha_i$ and $\gamma_{ij}$ are used in equation (2.44) to obtain an aggregate price index for energy. To do this the parameter $\alpha_0$ in equation (2.44) is determined so that the price of energy is equal to 1 in the U.S. in 1970. An energy price index is then calculated for each country.

Next we estimate the factor share equations (2.40), with $i$ and $j$ equal to capital, labor, and energy. In estimating these equations, we use our estimated aggregate price index for energy as an instrumental variable. We estimate these equations in stages, imposing additional parameter restrictions at each stage and testing each set of restrictions. In the first stage we impose only the restrictions implied by neoclassical production theory, that is, $\Sigma_i\, \alpha_i = 1$, $\Sigma_i\, \gamma_{Qi} = 0$, $\gamma_{ij} = \gamma_{ji}$, and $\Sigma_i\, \gamma_{ij} = \Sigma_j\, \gamma_{ij} = 0$. Next we add the homotheticity restrictions $\gamma_{Qi} = 0$. Finally, we test the restrictions that $\gamma_{ij} = 0$, such that the elasticities of substitution among all three factors are equal to 1.[33]

**Dynamic Versions of the Translog Cost Function**

A problem with the translog cost functions is that they do not describe differences between short-run and long-run elasticities, or how the adjustment to the long-run takes place. It is reasonable, for example, to expect that in the long run the aggregate production function exhibits constant returns to scale, but that it exhibits nonconstant returns in the short run. Although the limited scope of this study has not permitted us to estimate dynamic versions of the translog cost function, here we discuss ways of specifying such a function that permits adjustment to constant returns over time.

Constant returns requires that $\alpha_Q = 1$, $\gamma_{QQ} = 0$, and $\gamma_{Qi} = 0$ in equation (2.39). One way, then, to build in an adjustment to constant returns in the long run is to make these parameters functions of changes in output or prices. For example, the parameters $\alpha_Q$ and $\gamma_{QQ}$ could be specified as

$$\alpha_Q = \exp\left[\theta_1 \sum_{k=1}^{K} \lambda_1^k (\Delta Q_{t-k})^2\right] \tag{2.58}$$

and $$\gamma_{QQ} = \theta_2 \sum_{k=1}^{K} \lambda_2^k (\Delta Q_{t-k})^2 + \theta_3 . \tag{2.59}$$

Here the parameters $\theta_1$, $\theta_2$, and $\theta_3$ are estimated. If $\theta_3$ is not equal to zero, then nonconstant returns can exist even in the long run. Thus, if long-run constant returns is taken as given a priori, the estimate of $\theta_3$ (that is, whether it is significantly different from 0) provides a test of the correctness of the specification of the lag distribution. The lag distribution parameters $\lambda_1^k$ and $\lambda_2^k$ could be estimated if the data permitted, or they could be specified a priori (perhaps declining linearly). In either case $K$ might be three to five years. Of course, estimation of the parameters in (2.58) and (2.59)

33. We will also test restrictions pertaining to the characteristics of intercountry differences in the cost function, which involves the use of dummy variables that permit some of the parameters of the translog cost function to vary across countries. This is discussed later.

requires that the cost function (2.39) be estimated simultaneously with the share equations (2.40).

The parameters of $\gamma_{Qi}$ can also be made to adjust to zero in the long-run equilibrium by making them functions of a distributed lag in changes in prices:

$$\gamma_{Qi} = \theta_{Qi} \sum_{k=1}^{K} \lambda_{Qi}^{k}(\Delta \log P_{t-k})^2. \tag{2.60}$$

Thus, if the data do not permit the simultaneous estimation of the cost function with the share equations, $\alpha_Q$ and $\gamma_{QQ}$ could be assumed equal to 1 and 0 respectively, and adjustments could occur through the $\gamma_{Qi}$. In this case the production structure would be homothetic in the long run.

An alternative approach is to introduce the dynamic adjustment directly into the share equations. As we saw in our discussion of dynamic translog models for residential energy demand, this can be done by assuming that the shares adjust to a set of desired shares:

$$S_{i,t} = S_{i,t-1} + \sum_{j} \delta_{ij}(S_{j,t}^{*} - S_{i,t-1}), \tag{2.61}$$

where $S_{j,t}^{*}$ is given by equation (2.40). As before, adding up requires that the sum of all changes in shares be zero, so that

$$\sum_{i} \sum_{j} \delta_{ij}(S_{jt}^{*} - S_{j,t-1}) = \sum_{j} (S_{j,t}^{*} \sum_{i} \delta_{ij} - S_{j,t-1} \sum_{i} \delta_{ij}) = 0. \tag{2.62}$$

Since the $S_{j,t}^{*}$ and $S_{j,t-1}$ sum to 1, this equation implies the necessary condition that all of the columns of the matrix $(\delta_{ij})$ sum to the same arbitrary constant. Again if the number of shares is greater than 2, there are alternative constraints on the $\delta_{ij}$ that can be imposed to satisfy this condition.

**Multinomial Logit Models for Fuel Choice**

We saw earlier how the multinomial logit model can be used as an alternative means of studying the breakdown of energy consumption into demands for individual fuels in the residential sector. The logit model can be applied in the same way to the industrial sector. Although the logit model is not based on assumptions of cost minimization, it has properties that make it appealing for studying interfuel substitution. The model is consistent in that shares add to

1, shares respond to price changes in a way that is intuitively appealing, and the model is easy to estimate, which easily permits us to introduce alternative dynamic specifications.

We follow the same approach as for the residential sector: the logit model for four fuels is written as in equation (2.31), so that the relative shares of any two fuels can be represented by equation (2.32). Again only three equations are estimated because the parameters of the fourth equation are determined from the adding up constraint.

In estimating fuel shares, we include as attributes the relative price of each fuel, that is, the ratio of the real price of the fuel to the real price of energy, the latter being measured by the translog energy price aggregator described earlier. We do not include total energy expenditures in the industrial sector as an attribute since we a priori impose homotheticity on the fuel-share model. Other attributes, however, can include lagged quantity or share variables that allow shares to adjust dynamically to changes in price. Functional forms for the $f_i$ are somewhat arbitrary, but in a simple dynamic model they would be linear functions of the relative fuel prices $\tilde{P}_i = P_i/P_E$, where $P_E$ is the aggregate price of energy, output $Q$, and the lagged share $S_{i,t-1}$:

$$f_i(\mathbf{x}\boldsymbol{\beta}) = a_i + b_i\tilde{P}_i + c_iQ + d_iS_{i,t-1}. \tag{2.63}$$

The three estimating equations are then

$$\log (S_i/S_4) = (a_i - a_4) + b_i\tilde{P}_i - b_4\tilde{P}_4 + (c_i - c_4)Q + d_iS_{i,t-1} - d_4S_{4,t-1}, \qquad i = 1, 2, 3. \tag{2.64}$$

Note that two lagged shares appear in each equation. The three equations must again be estimated simultaneously, with both $b_4$ and $d_4$ constrained to be the same in each equation.

The results of estimating the static and dynamic versions of this model are discussed in the appendix to chapter 5. (Unfortunately, as in the residential sector, the model is not successful in providing statistically significant estimates of elasticities or dynamic adjustment times.)

## 2.3 Energy Demand in the Transportation Sector

The transportation sector accounts for a large proportion of total energy use, and in many countries has been the target of policies designed to reduce, or at least limit, the growth of energy consumption. In the U.S., for example, where gasoline alone accounted for some 17 percent of the total energy consumed in 1977, gasoline taxes, fuel efficiency standards for automobiles, taxes on crude oil, and oil price deregulation (which would lead to higher gasoline prices) have been debated as policy options. Much of this debate, however, has elicited widely differing views on the price responsiveness of gasoline demand. Clearly better estimates of the long-run price elasticity of gasoline demand, as well as the speed of adjustment of that elasticity, are needed for policy formulation.

Energy demand in the transportation sector is highly dependent on the existing stocks of fuel-burning vehicles (whether they be cars, trucks, or airplanes), the use of those stocks, and their average fuel-burning efficiency. (This last variable may be particularly important since many consumers have little flexibility in the use of their cars, and can achieve major savings in gasoline use only by purchasing smaller, more fuel-efficient ones.) We have constructed a model of the demand for motor gasoline that accounts for all of these variables, and we specify that model here. However, limited data availability did not permit the construction of such a model for other fuels used in the transportation sector, such as diesel fuel, aviation gasoline, and jet fuel. For these other fuels, and for fuels used in other sectors of energy use, we have been able to estimate only simple linear logarithmic models of demand. We turn now to the demand for motor gasoline, and in the appendix to this chapter we discuss the use of logarithmic models for other fuels.

The conventional wisdom in the U.S. has been that price elasticities of motor gasoline demand are small, perhaps −0.2 or −0.3, even in the long run, and this has been supported by some econometric studies using U.S. data. For example, Houthakker and Taylor (92) found price elasticities insignificantly different from zero, and Houthakker, Verleger, and Sheehan (93) found short-run and long-run elasticities of −0.075 and −0.24, respectively.

In fact, the true price elasticities may be much larger. Ramsey, Rasche, and Allen (144), working with a model of residential and commercial demand for gasoline that is based on a functional demand specification developed by Ramsey (143) and that includes price indices for competing forms of transportation, found a (long-run) own-price elasticity of around −0.7. Adams, Graham, and

Griffen (1) argued that the use of data for a single country like the U.S., where prices have varied over only a small range, would bias elasticity estimates downwards. They used cross-sectional data covering twenty OECD countries for the year 1969 to estimate a multi-equation model describing gasoline consumption per car, the average weight of cars, average horsepower of cars, and the stock of cars per capita, and found an own-price elasticity for gasoline demand of −0.9.

As we have explained in the first chapter of this book, long-run elasticities can best be identified by pooling data over a number of countries. In addition, the demand for gasoline is best treated as a derived demand, that is, dependent on the stock of cars and the average size and horsepower (or fuel efficiency) of this stock, as well as on price and income directly. We have therefore specified a model of gasoline demand along the lines of the Adams, Graham, and Griffin model, where the stock of cars and characteristics of cars are endogenous (and dependent in part on the price of gasoline), and in turn are explanatory variables of gasoline demand.

The model is based on the following identity for the consumption of gasoline $Q$:

$$Q = \frac{STK \cdot TVPC}{EFF} \tag{2.65}$$

where $STK$ is the stock of automobiles, TVPC is traffic volume per car (that is, the average number of kilometers driven per car each year), and $EFF$ is the average gasoline-burning efficiency of the stock of cars in miles per gallon. Three equations describe the stock of cars; the first explains additions to the stock, the second explains the rate of depreciation of the stock, and the third is an accounting identity. Next, an equation is specified for traffic volume, and a last equation describes the average efficiency of the stock of cars.

The dynamic response of gasoline consumption to changes in the price of gasoline will be a function of the responses of each of the variables on the right-hand side of equation (2.65). Generally we would expect the stock of cars and the efficiency of the stock to respond only slowly to price changes, while traffic volume may respond more quickly. Thus, even if each of these variables responds to price with a geometric (Koyck) lag, consumption itself may have a more complicated lag structure that must be determined by simulating the full estimated model.

Because we model gasoline consumption as the product of a number of different variables, each of which is a dynamic function of other explanatory variables, we do not derive our model from a generalized utility function or cost function as we did for the translog models of residential and industrial energy demand. Instead, we describe composite variables of equation (2.65) as linear or linear-logarithmic functions of particular explanatory variables.

We begin with the stock of passenger cars. This variable itself is not sensitive in the short run to changes in the price of cars, the price of gasoline, or income; instead, only additions to the stock (both net and replacements of worn-out cars) and, to a lesser extent, the rate of depreciation of the stock will be sensitive to these variables. The stock itself is given by an accounting identity:

$$STK_t = (1 - r)STK_{t-1} + NR_t, \tag{2.66}$$

where $r$ is the rate of depreciation of the stock, and $NR$ is new registrations (that is, total additions to the stock).

New registrations can be thought of as bringing the actual stock of cars closer to the desired stock, where the desired stock is a function of explanatory variables such as the price of cars, $P_i$, the price of gasoline, $P_g$, and income $Y$. Thus, if per capita new registrations depends on the per capita desired stock ($STK^*/POP$) according to

$$\frac{NR_t}{POP_t} = w\left(\frac{STK_t^*}{POP_t} - \frac{STK_{t-1}}{POP_{t-1}}\right) + r\frac{STK_{t-1}}{POP_{t-1}} + \lambda\frac{NR_{t-1}}{POP_{t-1}},$$

and if $STK^*$ is a linear function of $P_c$, $P_g$, and $Y$, $NR$ will be given by

$$\frac{NR_t}{POP_t} = a_0 + a_1P_c + a_2P_g + \frac{a_3Y}{POP} - (w - r)\frac{STK_{t-1}}{POP_{t-1}} + \frac{\lambda NR_{t-1}}{POP_{t-1}}. \tag{2.67}$$

Note that in this equation the speed of adjustment of the actual to the desired stock varies over time; the annual speed of adjustment is $w$ in the short run, and $w/(1 - \lambda)$ in the long run.[34]

The depreciation rate $r$ will vary across countries, and is also likely to be a function of prices and income. We would expect rising per capita incomes to increase the depreciation rate (as consumers are better able to afford newer cars), and a higher price of cars to decrease the rate. We estimate a linear equation of the form

$$r = \sum_i b_{0i} D_i + \frac{b_1 Y}{POP} + b_2 P_c, \tag{2.68}$$

where the $D_i$ are country dummy variables.

Equations (2.66), (2.67), and (2.68) together determine the stock of cars. The next equation describes the average number of kilometers driven per car, in other words, traffic volume per car TVPC. This variable should exhibit a strong positive dependence on per capita income, and may also depend negatively on the price of gasoline. The dependence on both variables should occur with a lag, so we include the lagged dependent variable on the right-hand side of a log-linear equation:

$$\log TVPC = \sum_i c_{0i} D_i + c_1 \log \left(\frac{Y}{POP}\right) + c_2 \log P_g + c_3 \log TVPC_{t-1}. \tag{2.69}$$

The last equation describes the average fuel efficiency of the stock of cars. Average efficiency should respond to changes in the price of gasoline, but only with a long lag, and its mean value may also vary considerably across countries. We estimate a log-linear equation for the inverse of efficiency:

$$\log \left(\frac{1}{EFF}\right) = \sum_i d_{0i} D_i + d_1 \log P_g + d_2 \log \left(\frac{1}{EFF_{t-1}}\right) \cdot \tag{2.70}$$

34. A similar model describing expenditures on automobiles in the U.S. has been constructed by Hymans (97) using quarterly data. In his model, however, the depreciation rate is fixed, and there is no lagged dependent variable in the equation corresponding to (2.67).

One problem with this equation is that it imposed no limit on efficiency; if the price of gasoline rises sufficiently, average efficiency can grow beyond any physically plausible value. For this reason we also estimate an alternative version of this equation:

$$\log\left(\frac{1}{EFF - K}\right) = \sum_i d'_{0i}D_i + d'_1 \log P_g + d'_2 \log\left(\frac{1}{EFF_{t-1} - K}\right), \qquad (2.70a)$$

where $K = 1/EFF_{\max}$ defines the upper limit on average efficiency.

Since we view per capita income and the prices of automobiles and gasoline as exogenous variables, we can obtain consistent parameter estimates by estimating these equations individually. However, more efficient estimates can be obtained by treating the equations as a set and using Zellner estimation to account for possible cross-equation error correlations.

## Appendix: Demand Models for Other Fuels and Sectors of Use

In estimating the demands for other fuels and for energy use in other sectors, we use simple static and dynamic linear-logarithmic equations. In chapter 7, these models are also applied to the estimation of demands for particular oil products in some of the larger (in terms of energy consumption) developing countries. We will find, however, that although these models are simple, in some cases they yield robust estimates of demand elasticities.

Since we are working with pooled time-series cross-section data, it is sometimes difficult to identify separately short-term and long-term effects, and determine the relative contributions to each from the cross-section versus time-series variation in the data. In a 1965 study of the breakdown of total consumption expenditures, Houthakker (91) separated short-run and long-run elasticities by running separate regressions across countries and across time. We can apply this approach to both the static and dynamic versions of linear-logarithmic demand models.

The basic demand equation for fuel $i$ is

$$\log q_{ijt} = \alpha_i + \beta_i \log y_{jt} + \gamma_i \log P_{ijt} + \delta_i t + \epsilon_{ijt}, \qquad (2.71)$$

where $j$ is the country index and $t$ the time index, $q_i$ is per capita consumption of fuel $i$, $y$ is per capita income in real terms, and $p_i$ is the relative price of fuel $i$ (that is, relative to the prices of all fuels). These equations can be estimated using ordinary least squares or simple weighted least squares.

Short-run and long-run effects can be identified by estimating "within country" and "between country" regressions. The within country (short-run) regression is

$$\log q_{ijt} - \overline{\log q_{ij}} = \beta_i(\log y_{jt} - \overline{\log y_j}) + \gamma_i(\log p_{ijt} - \overline{\log p_{ij}}) , \tag{2.72}$$

and the between country (long-run) regression is

$$\overline{\log q_{ij}} = \beta_i \overline{\log y_j} + \gamma_i \overline{\log p_{ij}} + \delta_i . \tag{2.73}$$

Here the bar represents averaging over time. Note that the within country regression is pooled, while the between country is purely cross sectional, and that deviations from means (over time) are used in the within country regressions in order to eliminate long-run effects. Equation (2.72) can also be run for each country separately, in order to determine how elasticities vary across countries. Alternatively, a pooled regression can be performed, with a multiplicative district dummy variable introduced to one coefficient at a time; for example,

$$\log q_{ijt} - \overline{\log q_{ij}} = \sum_j \beta_{ij} D_j(\log y_{jt} - \overline{\log y_j}) + \gamma_i(\log p_{ijt} - \overline{\log p_{ij}}) . \tag{2.74}$$

Equations (2.72) and (2.73) can also be estimated in first-differenced form. This crudely reduces trend effects, and also eliminates problems associated with the arbitrary choice of purchasing power parities. The within country version is

$$\Delta \log q_{ijt} - \overline{\Delta \log q_{ij}} = \beta_i(\Delta \log y_{jt} - \overline{\Delta \log y_j}) + \gamma_i(\Delta \log p_{ijt} - \overline{\Delta \log p_{ij}}) , \tag{2.75}$$

and the between country version is

$$\overline{\Delta \log q_{ij}} = \beta_i \overline{\Delta \log y_j} + \gamma_i \overline{\Delta \log p_{ij}} . \tag{2.76}$$

We can also specify a dynamic version of (2.72). We assume that demand $q_{it}$ depends not only on price and income in period $t$ but also on a state variable, $s_{it}$:

$$\log q_{it} = a_i + b_i \log y_t + c_i \log p_{it} + d_i \log s_{it}, \tag{2.77}$$

where $s_{it}$ may represent a stock or a habit level. The dynamics of $s_{it}$ can be expressed as

$$\Delta \log s_{it} = \log q_{it} - w_i \log s_{i,t-1}, \tag{2.78}$$

where $w_i$ is effectively a depreciation rate. To obtain the demand equation, rewrite equation (2.77) as

$$s_{it} = \frac{1}{d_i} (\log q_{it} - a_i - b_i \log y_t - c_i \log p_{it}), \tag{2.79}$$

which can be substituted into equation (2.78) such that

$$\begin{aligned} \Delta \log s_{it} = \log q_{it} - \frac{w_i}{d_i} (\log q_{i,t-1} - a_i \\ - b_i \log y_{t-1} - c_i \log p_{i,t-1}). \end{aligned} \tag{2.80}$$

Now first-difference equation (2.79) and substitute into (2.80) to yield the estimating equation

$$\begin{aligned} \log q_{it} = \alpha_0 + \alpha_1 \log q_{i,t-1} - \alpha_2 \Delta \log y_t + \alpha_3 \log y_{t-1} \\ - \alpha_4 \Delta \log p_{it} + \alpha_5 \log p_{i,t-1}. \end{aligned} \tag{2.81}$$

This equation can also be estimated within countries and between countries. If this results in differences in the estimated value of $\alpha_1$, it would indicate that the adjustment response is not constant over time.

# 3 The Estimation of Energy Demand Models

The models of sectoral energy demands developed in the last chapter are estimated using data spanning a number of industrialized countries. The use of international data is an important aspect of this study in that it enables us to determine the long-run characteristics of energy demand and to estimate long-run price and income elasticities. We present the statistical results from the estimation of these models in chapters 4, 5, and 6. First, however, it is necessary to resolve a number of methodological issues involved in the estimation itself. We discuss these issues in the first part of this chapter. In the second part of the chapter we briefly describe the sources and characteristics of all of the data used in the study.

## 3.1 Methodological Issues in the Estimation of Demand Models

A number of problems must be considered in estimating the models described in the last chapter, which arise from the fact that we are using international data for estimation and from the nature of the models themselves. The first problem comes about because the comparison of expenditures or prices in different countries requires valuing different currencies in terms of a common unit. As we will see, the use of purchasing power parities for this purpose is probably more desirable than the use of exchange rates, but the choice of the particular purchasing power parity index is not always clear. The second problem involves the choice of whether to value energy quantities in gross or net terms, that is, whether to adjust for the different thermal efficiencies of different fuels. Third, for both the residential and industrial sectors, energy price indices must be obtained that aggregate the prices of the individual fuels used in each sector, and there are alternative approaches for obtaining such indices. A fourth problem has to do with the identification of inter-country differences in the structure of consumption or production. Finally, there are a number of issues involving the choice of econometric method for the estimation of our models. We deal with each of these problems in turn here.

### Use of Purchasing Power Parities

Since all of the price and expenditure data for each country in our sample are measured in terms of the local currency of that country, estimation with data covering the pooled sample requires a method to convert these numbers into common units. One method that has been used in a number of studies in the past is simply to use official exchange rates.[1] This can be misleading, however, since official rates can differ considerably from the exchange rates that would have prevailed under free market equilibrium, and the tariffs, quotas, subsidies, and other controls in effect in many countries can result in price structures that differ considerably from relative international prices.

Alternatively, we could attempt to identify "free market" exchange rates between individual countries over time periods thought to reflect equilibrium conditions, for example, over periods during which trade balances were near zero. Even under free trade, however, equilibrium exchange rates only reflect the price equalization of internationally traded goods, which for most countries represents a small subset of all market goods. As Chenery and Syrquin (29) have pointed out, the relative price of nontraded goods can be expected to increase with real per capita income, so that the use of official (or free market) exchange rates can lead to an underestimate of the purchasing power of the currencies of lower income countries.

A better approach is to use purchasing power parity indices (PPP) to convert national currencies to some base currency. Purchasing power parities can be obtained explicitly by making binary comparisons between a base country (say, the U.S.) and various other countries, using a fixed set of quantity weights.[2] The problem, of course, is that two sets of price index numbers (Laspeyre and Paasche) can be obtained depending on whether base country or other country weights are used. In this work we use a Fisher "ideal" index (a geometric mean of these two index numbers) as a

1. This approach has been used by Adams and Griffin (2) and Goldberger and Gameletsos (66).

2. Purchasing power parities can also be obtained implicitly by dividing a nominal national currency estimate of national product (or one of its components) by a base currency estimate of the same national product. This procedure was used recently by Lluch and Powell (112). For a general discussion of explicit purchasing power parities, see Belassa (7) and Allen (5).

single index of relative purchasing power. The use of a Fisher ideal index is suggested on theoretical grounds by Samuelson (147) and is supported on empirical grounds by Kloek and Theil (106).

In our work on energy demand in the residential sector we use purchasing power parities by consumption category calculated by the German Statistical Office (150). These indices were obtained by means of detailed price comparisons.[3] They are binary index numbers for which Germany is the base country, and we must therefore use Germany as a bridge to convert to the U.S. as base country. It is important to note that although binary PPP permit us to make a transitive international ordering of purchasing powers, this ordering is not invariant with respect to the choice of bridge country. Kravis et al. (108) also calculated multilateral PPP by means of a regression model that estimates the purchasing power parity for a single category of expenditure as a function of all other international price ratios. Unfortunately, some of the countries in our sample are not included in the Kravis study.

The resulting parities apply to a particular base year, but we must construct intertemporal indices to deflate our time series; in other words, to obtain for each country relative prices over time for each category of consumption. We do this using implicit price indices for each consumption category in each country, thus constructing an implicit ratio of relative intertemporal purchasing power in terms of a base-year numeraire (normalized so that 1970 is our base year). The resulting base-year purchasing power parities for each consumption category are shown in table 3.1.[4] Note that for purposes of estimating our consumption model these purchasing power parities provide a set of base-year relative prices for each consumption category.

Purchasing power parity indices are also used in estimating energy demand models for the industrial and transportation sectors. Indices for the total gross domestic product of each country are needed for both the industrial and transportation sectors, while our models of

3. Binary purchasing power parities were more recently calculated by Kravis et al. (108) but for only a subset of the countries in our sample.
4. For those countries also covered by Kravis et al., our numbers are at all times within 10 percent of the 1970 Kravis numbers.

the industrial sector also require indices for investment goods. Here we are able to use purchasing power parities with the U.S. as base country, so that bridging across a third country is not necessary. Our indices come from a number of different sources, and we discuss these later in this chapter when we describe all of the data used for the study.

**Gross vs. Net Energy Consumption**

As pointed out by Adams and Miovic (3), alternative fuels are not equivalent on a calorific basis as a result of the differing thermal efficiencies of energy-consuming equipment. If more efficient fuels are substituted for less efficient fuels over time, the measurement of an overall energy elasticity (that is, the percent change in energy use associated with a 1 percent change in GNP) will yield a larger number if thermal efficiencies are taken into account. Adams and Miovic, in fact, estimate an overall energy elasticity for the U.S. and several European countries of about 0.8 when gross energy quantities are used, and about 1.0 when energy quantities are adjusted for thermal efficiencies.

This has led some individuals to suggest the use of net energy consumption (adjusted for thermal efficiencies) rather than gross energy consumption in the estimation of energy demand models. For example, in recent studies of energy demand Nordhaus (123), Adams and Griffin (2), and Fuss and Waverman (62), made the assumption that within each sector fuels are perfect substitutes, so that (given equal levels of nonfuel cost) interfuel competition is determined by relative net prices of fuels. Net consumption and net price are thus given by

$$QN_{ij} = \eta_{ij} Q_{ij} \tag{3.1}$$

and

$$PN_{ij} = \frac{P_{ij}}{\eta_{ij}}, \tag{3.2}$$

where $\eta_{ij}$ is the efficiency of fuel $i$ in sector $j$.

There are two problems with this approach. First, it is difficult to obtain reliable estimates of thermal efficiencies. Identification problems make econometric estimates infeasible unless unduly restrictive structural assumptions are imposed, and engineering estimates differ considerably from source to source.[5] As an example of this problem, we show in table 3.2 engineering estimates of thermal efficiencies cited or used in four different studies. Note that these estimates differ considerably from study to study, and in fact we have no way of determining what a correct estimate should be.

A second and more fundamental problem is that fuels are not perfect substitutes (particularly in the short run), and there are nonthermal efficiencies (which we could label "economic") that also affect consumer demand. Fuel choice is based not only on usable thermal content but also on such factors as convenience, controllability, cleanliness, and capital costs, and the effects of these efficiencies (as well as thermal ones) would hopefully all be manifested in the estimated parameters of our demand models. It thus does not seem particularly relevant to measure fuel consumption in efficiency-adjusted thermal units, any more than it would be to measure food consumption in net calorific terms.[6]

We therefore choose to measure all of our energy quantities in gross rather than net terms. We assume that both thermal and nonthermal efficiencies have effects on interfuel competition, and that these effects will be picked up in the way that estimated fuel expenditure shares change as relative prices and income change.

**Price Indices for Aggregate Energy Use**

Our model of residential energy use begins with the breakdown (using the indirect translog utility function) of total consumption expenditures into six different consumption categories, of which

5. Adams and Miovic (3) attempted to measure thermal efficiencies by assuming that fuel inputs are a constant proportion of aggregate economic output and that there is no substitution between fuel inputs and labor and capital. Their production function was thus

$$Y = \min(\alpha F, f(L, K)),$$

where fuel input $F$ is given by $F = \Sigma_i \eta_i h_i F_i$, where $h_i$ is the calorific content of fuel $i$ ($F_i$). Since the $h_i$ are known, they can estimate the $\eta_i$ up to a scalar multiple. The assumptions are extremely restrictive, however, and their results differ considerably from engineering estimates that they cite. For an engineering discussion of thermal efficiencies, as well as a set of estimates, see Hottel and Howard (88).

6. This is discussed further by Turvey and Nobay (160).

energy is one. The estimation of this model therefore requires a price index for aggregate energy use (as opposed to prices for individual fuels). Similarly, our model of industrial energy use involves the breakdown of total costs of production into expenditure shares for capital, labor, and energy, so that again a price index is needed for aggregate energy use in this sector.

For the residential sector it would be possible to construct an implicit price index from nominal and real energy expenditure series. However, since price series for individual fuels are available, it would be preferable to use this data directly. But a price index that truly reflects the unit cost of energy will not equal a simple weighted average of fuel prices because fuels are not perfect substitutes.

A typical approach is to construct an approximate Divisia index as a means of aggregation.[7] An alternative approach is to specify (and estimate) an aggregator function that relates the aggregate price index to the component prices. Any unit cost function could be used to represent the aggregate price of energy, but a logical choice is the translog cost function.[8] As an incidental advantage, the translog cost function (or aggregator) provides us with an instrumental variable for the price of energy when estimating the aggregate consumption model.

The translog cost function (which is equivalent to a homothetic and stationary indirect utility function with unit total expenditure) is given by

$$\log P_E = \gamma_0 + \sum_i \gamma_i \log P_i + \sum_i \sum_j \gamma_{ij} \log P_i \log P_j. \tag{3.3}$$

Assuming cost-minimizing behavior, the fuel-share equations are then

$$S_i = \gamma_i + \sum_j \gamma_{ij} \log P_j, \qquad i = 1, \ldots, n. \tag{3.4}$$

7. See Jorgenson and Griliches (99) and Hulten (95).

8. This is appealing as an unrestrictive representation of unit cost. As Diewert (54) has shown, the Divisia index is exact for the translog cost aggregator function, that is, it retrieves the actual values of the translog cost function. However, our translog cost aggregator is an approximate index since it is an approximation to the true cost function (and there is therefore no correspondence to an exact Divisia index).

The first $(n - 1)$ share equations are estimated subject to the restrictions $\Sigma\gamma_i = 1$, $\gamma_{ij} = \gamma_{ji}$, and $\Sigma_i\gamma_{ij} = 0$. The estimated parameters $\gamma_i$ and $\gamma_{ij}$ are then substituted in equation (3.3). Using data for the fuel prices, $P_i$, the price index $\hat{P}_E$ can be computed. Note that the energy price index $\hat{P}_E$ is determined only up to an unknown scalar multiple $\gamma_0$. The procedure is to pick one country (say, the U.S.) as a base country, and then solve equation (3.3) for $\gamma_0$ so that the price of energy in the base country is equal to 1 in some base year (say, 1970). Relative price indices are thereby determined for all years in all of the other countries.

A problem remains regarding the number of fuels to be included in equations (3.3) and (3.4). Although four fuels are included in our demand model, almost no coal is consumed in the residential sectors of the U.S. and Canada. This suggests that equations (3.3) and (3.4) should apply to a three-fuel aggregation (oil, gas, and electricity) for the U.S. and Canada, and a four-fuel aggregation for the remaining countries. Should this approach be used—as opposed to a four-fuel aggregation for all countries—a method is needed to bridge the U.S.-Canadian aggregator with the aggregator for the remaining countries. The following bridging method could be used:

1. Equation (3.4) is estimated for four fuels for all countries except the U.S. and Canada. The unidentified parameter $\gamma_0$ in equation (3.3) is chosen so that the price of energy is equal to 1 in one of the European countries, say, Belgium, in 1970. This permits the calculation of the price of energy for all countries except the U.S. and Canada relative to Belgium in 1970.

2. Equation (3.4) is estimated for four fuels for all nine countries. The parameter $\gamma_0$ is chosen so that the price of energy is equal to 1 in the U.S. in 1970. The resulting index gives us the price of energy in Belgium in 1970 relative to that in the U.S. in 1970.

3. Equation (3.4) is then estimated for three fuels (oil, gas, and electricity) for the U.S. and Canada only. The parameter $\gamma_0$ is chosen so that the price of energy is equal to 1 in the U.S. in 1970. Now, using the Belgium-to-U.S. conversion ratio determined in step 2, the price indices calculated in step 1 are converted to a U.S. 1970 = 1 base.

Because the quantities of coal consumed in the residential sectors of the U.S. and Canada are small but not zero, it is not clear whether the bridging approach described above or a simple four-fuel aggregation across all countries should be used to construct the energy price index. We therefore estimate energy price indices using both methods. If the resulting indices are nearly the same, this would indicate that the relative size of the coal shares in the U.S. and Canada do not distort the fit of a fuel choice model that includes four fuels for all countries. In this case coal could be included for all countries in the fuel demand models. If the results are significantly different, then coal should not be included in the U.S. and Canada demand models, and the bridging method should be used to calculate the energy price index.

Note that this problem does not exist for our industrial demand model. Substantial quantities of all four fuels are used in the industrial sectors of every country in our sample, so that equation (3.4) is estimated for all of the fuels.

**Identifying Intercountry Differences in Elasticities**

One of our objectives in this study is to determine if and where energy demand elasticities vary across countries, and the possible reasons for such variation. To identify regional variations in elasticities, we must specify alternative ways of allowing for regional parameter variation when our models are estimated with pooled data.

At the one extreme, we might assume that the parameters of our models are the same for all countries. For example, in the residential sector, where the indirect translog utility function is used, the share equations could be estimated by simply pooling all of the data together. This, of course, would restrict the parameters $\alpha_j$, $\beta_{ji}$, and $\beta_{jt}$, to be the same in each country. The resulting price elasticities would still vary across countries, but only because relative prices and total expenditures are different in different countries. At the other extreme we could estimate our models for each country separately; in the translog case $\alpha_j$, $\beta_{ji}$, and $\beta_{jt}$ could be different for every country. While this specification is least restrictive, it is likely to be infeasible due to insufficient data, or to result in short-run elasticity estimates.

There are two compromise approaches that could be followed. One is to estimate models by pooling subsets of countries, so that parameters can differ across subsets but are the same within each subset. This might involve, for example, pooling the U.S. and Canada separately from some or all of the European countries. Re-estimation using alternative groupings could then be used to determine the validity of constraining parameters to be the same across countries in a subset.

A second approach is to pool all of the countries, but to introduce regional dummy variables that would allow a subset of a model's parameters to vary across countries. In the case of the indirect translog utility function, we might assume that the coefficients $\alpha_j$ of the first-order terms in the Taylor series approximation can vary across countries, while the coefficients $\beta_{ji}$ and $\beta_{jt}$ are the same for each country. This would involve estimating the following share equations:

$$S_j = \frac{\sum_k \alpha_{jk} D_k + \sum_i \beta_{ji} \log (P_i/M) + \beta_{jt} \cdot t}{\sum_k \alpha_{Mk} D_k + \sum_i \beta_{Mi} \log (P_i/M) + \beta_{Mt} \cdot t},$$

$$j = 1, \ldots, (n - 1), \tag{3.5}$$

where $D_k$ are country dummy variables ($D_k = 1$ for country $k$, and 0 otherwise). Note that the usual restrictions on the $\beta_{ji}$ and $\beta_{jt}$ apply, but $\alpha_{Mk} = \Sigma_j \alpha_{jk} = -1$ for each country $k$. For the case of the translog cost function, as applied to our industrial energy demand model, the first-order coefficients $\alpha_i$ would again vary across countries, while the coefficients $\gamma_{Qi}$ and $\gamma_{ij}$ would be the same for each country. This would mean estimating the following share equations:

$$S_i = \sum_k \alpha_{ik} D_k + \gamma_{Qi} \log Q + \sum_j \gamma_{ij} \log P_j, \tag{3.6}$$

where $D_k$ are country dummy variables ($D_k = 1$ for country $k$ and 0 otherwise). The usual restrictions on the $\gamma_{ij}$ and the $\gamma_{Qi}$ again apply, but $\Sigma_i \alpha_{ik} = 1$ for each country $k$. As we will see, an advantage of this method is that it partially deals with heteroscedasticity of the error terms within each equation; it is essentially the covariance method for estimation with pooled data.

Alternatively, we could assume that the coefficients of the second-order terms can vary across countries, while the coefficients of the first-order terms are the same for each country. For the residential sector, the share equations from the indirect translog utility function would then be

$$S_j = \frac{\alpha_j + \sum_k \sum_i \beta_{jik} D_k \log (P_i/M) + \beta_{jt} \cdot t}{\alpha_M + \sum_k \sum_i \beta_{Mik} D_k \log (P_i/M) + \beta_{Mt} \cdot t},$$

$$j = 1, \ldots, (n - 1). \tag{3.7}$$

Note that the restrictions on $\beta_{jik}$ are now that $\beta_{jik} = \beta_{ijk}$ for each country $k$, and that $\beta_{Mik}$ is the same in each share equation for every country $k$. For the industrial sector, the share equations from the translog cost function are now written as

$$S_i = \alpha_i + \gamma_{Qi} \log Q + \sum_k \sum_j \gamma_{ijk} D_k \log P_j, \tag{3.8}$$

with the restrictions that $\gamma_{ijk} = \gamma_{jik}$ and $\Sigma_i \gamma_{ijk} = \Sigma_j \gamma_{ijk} = 0$ for every country $k$.

Finally, note that variables that vary only regionally, or that vary both regionally and timewise, can be introduced in addition to the regional dummy variables. In the case of the indirect translog utility function, for example, we might assume that $\alpha_j$ is a function of temperature $T$ (which has both a regional and timewise variation),

$$\alpha_j = a_j + b_j T, \tag{3.9}$$

with $a_j$ varying across countries, so that the share equations are

$$S_j = \frac{\sum_k a_{jk} D_k + b_j T + \sum_i \beta_{ji} \log (P_i/M) + \beta_{jt} \cdot t}{\sum_k a_{Mk} D_k + \sum_i b_i T + \sum_i \beta_{Mi} \log (P_i/M) + \beta_{Mt} \cdot t}, \tag{3.10}$$

with $a_{Mk} = \Sigma_i a_{ik} = -1$ for each country $k$.

There is no a priori reason for preferring any of the above specifications (other than the econometric convenience of allowing the coefficients of the first-order terms to vary across countries). However, for either specification, the null hypothesis that the corresponding coefficients are the same across countries (in other words, that the indirect utility function or cost function is regionally homogeneous) can be tested using the straightforward chi-square test described in chapter 2.

In estimating demand models, we test the various approaches described above for introducing regional heterogeneity. We find that in most cases low-variance parameter estimates cannot be obtained if the second-order coefficients are allowed to vary across countries—there are simply too many parameters involved given the limited number of degrees of freedom available for estimation. On the other hand, we find that in every case the data strongly support regional variation of the first-order coefficients. Also, in some cases there is evidence for pooling one or more subsets of countries (usually the U.S. and Canada) separately.

**Estimation Methods**

The choice of econometric method to be used for the estimation of each model involves a trade-off between the richness of the stochastic specification (and hopefully a resulting gain in efficiency) and computational expense. This trade-off is particularly severe given that most of our models involve systems of equations with cross-equation parameter constraints. Ideally, we would like to estimate these equations under the assumption of a stochastic specification for which the error terms are heteroscedastic, autocorrelated both across time and across countries within each equation, and correlated across equations in the system. However, estimating such a specification (which amounts to full generalized least squares) would be unreasonably costly even if the individual equations were linear in the parameters. If individual equations are nonlinear in the parameters (as is the case with the share equations for the nonhomothetic version of the indirect translog utility function), the estimation might be computationally infeasible. We must therefore settle for a more restrictive specification that would hopefully capture the more important characteristics of the error terms.

Another argument in favor of a more restrictive and simpler stochastic specification is that this specification in any case is fairly arbitrary. The use of additive error terms on the share equations may itself be a misspecification. Usually it is argued that these errors are the results of errors in the cost minimization or utility maximization process. However, errors added to the first-order conditions of a cost minimization, for example, need not result in additive errors in the translog share equations. Thus there is an inherent degree of arbitrariness in the stochastic specification, and this makes the use of a more simple error structure appealing.

Our approach is to seek a balance between the possible gain in efficiency resulting from a richer stochastic specification and the corresponding increase in computational cost. Therefore, when estimating translog models (which can be nonlinear in the parameters and/or have cross-equation parameter constraints that are nonlinear), we ignore error term heteroscedasticity and autocorrelation within equations, and account only for error correlations across equations.[9] To do this we use iterative nonlinear Zellner estimation. This technique iterates over the cross-equation error covariance estimates, and, if the equations are nonlinear, simultaneously iterates over successive linearizations of the individual equations. Under the assumption of no heteroscedasticity or autocorrelation within equations, iterative Zellner estimation is equivalent to full-information, maximum-likelihood estimation.[10] However, we limit the number of iterations on the error covariance matrix to five; this reduces computational expense while still capturing at least 90 percent of the added efficiency that results from accounting for cross-equation error correlations.

As explained earlier, when estimating translog share equations a choice must be made as to whether the first-order and/or second-order coefficients in the translog approximation to the indirect utility function or cost function should be allowed to vary across countries. Allowing both sets of coefficients to vary across countries is equivalent to estimating a separate model for each country, but this would require more data than is available, and in addition is likely to result in short-run elasticity estimates. We have estimated translog share equations letting only the second-order coefficients vary across countries, but a large number of coefficients are involved (unless restrictions are imposed a priori on the structure of consumers' preferences or production), and again this leaves too few degrees of freedom to give satisfactory results. We therefore allow the first-order coefficients to vary across countries through the use

9. Accounting for within-equation heteroscedasticity and autocorrelation is certainly possible even if the equation is nonlinear in the parameters. One might use an algorithm that repeatedly linearized each equation, iteratively computed an error covariance matrix, and then estimated a transformed linear equation for each linearization; see, for example, Eisner and Pindyck (58). There is no guarantee, however, that final convergence will ever occur, but, even if it does, the process is likely to be extremely expensive.
10. See Zellner (165) and Gallant (63). Oberhofer and Kmenta (124) prove that iterative Zellner estimation (iterating to convergence on the cross-equation error covariances) is equivalent to full-information maximum likelihood.

of regional intercept dummy variables in the share equations. It turns out that this approach has an important additional advantage—it is equivalent to a covariance model of the implicit error terms in the share equations (that is, error terms composed of a regional component and a total component).[11]

The logit models and logarithmic models specified in the last chapter are all linear in the parameters, so that the stochastic specification can be somewhat richer. When estimating the equations of these models, we can also account for within-equation heteroscedasticity using the following procedure. First, each equation in the system is estimated using ordinary least squares. The resulting regression residuals, which we can label $u_{kt}$, can then be used to obtain consistent estimates of the regional (country) error variances $\sigma_k^2$:

$$\hat{\sigma}_k^2 = \frac{1}{T-m-1}\sum_t (u_{kt})^2, \tag{3.11}$$

where $T$ is the number of annual observations for country $k$ and $m$ is the number of independent variables in the equation. Different estimates of $\hat{\sigma}_k^2$ will be of course obtained for each equation in the system. The data can then be transformed by dividing each observation by the appropriate estimated error term standard deviation $\hat{\sigma}_k$, and the entire system of equations is re-estimated using iterative Zellner estimation.[12]

All of the estimation work in this study has been carried out using the GREMLIN experimental nonlinear estimation package, which is part of the TROLL econometric software system. This package permits the user to perform iterative nonlinear Zellner estimation conveniently and with reasonable computational expense.[13]

11. For a discussion of the covariance model in estimation with pooled data, as well as the related error components model, see Pindyck and Rubinfeld (140), and Wallace and Hussain (163).
12. MacAvoy and Pindyck (113) and Pindyck (137) used a similar approach to single-equation estimation that also accounted for timewise autocorrelation.
13. For details on the estimation algorithm and its use, see Belsley (13). For a discussion of alternative nonlinear estimation algorithms, see Berndt, Hall, Hall, and Hausman (19), Chow (31), and Gallant (63). The TROLL

## 3.2 The Data

We now turn to the construction of the data series used in this econometric study. As we explained here, the sources for our data varied, and in some cases data for a particular variable spanning a range of years and a range of countries had to be pieced together from more than one source. Only some of our data was obtained from such standard sources as the OECD or the U.N.'s statistical office because in many cases the needed data were not collected by these agencies, and in other cases the data had been collected but had been aggregated or categorized in ways limiting their usefulness for this study. As a result it was often necessary to go to the national statistical yearbooks of individual countries to obtain data.

All of the data used in this study are described in considerable detail in a data base "User's Guide" by J. Carson (28). (In addition, the MIT Energy Laboratory will make the data available directly to other researchers wishing to replicate or extend this study or conduct studies of their own.) Here we provide a summary of the sources and methods by which the data series were constructed. We describe the data used to estimate the translog models and logit models for the residential and industrial sectors, the data used for the model of gasoline demand, and the price and quantity data used to estimate the logarithmic equations of fuel demands.

### The Residential Sector

Our models of energy demand in the residential sector explain the breakdown of total household consumption expenditures into six categories of consumption, of which energy is one, and the breakdown of expenditures on energy into expenditures on each of four fuels. In order to estimate these models, data is needed for the real expenditures allocated to each category of consumption, relative price indices for each category of consumption, expenditures (or quantities) and prices for each fuel, and such variables as disposable income and population.

---

software system is maintained by the Center for Computational Research in Economics and Management Science at MIT.

Nine countries are included in our sample: Belgium, Canada, France, Italy, the Netherlands, Norway, the U.K., and U.S., and West Germany. The data collected for these countries are described briefly here:

*Consumption Expenditures* These are broken down into six categories—food (including alcohol and tobacco), clothing, durable goods, transportation and communication, energy, and "other." This last category includes housing expenditures (actual and imputed rental payments), expenditures on health services, and any other consumption expenditures. Data were obtained from the OECD's *National Accounts,* the national accounts publications of the EEC's statistical office, the U.N.'s *Yearbook of National Accounts,* and the national statistical yearbooks of individual countries. The data are measured in current local currency units but are converted into 1970 U.S. dollars, using purchasing power parity indices by consumption category.

*Price Indices for Consumption Expenditures* A retail price index (1970 = 100) was collected for each of the categories of consumption expenditures listed above. Although for some countries retail price indices were available directly, implicit price indices were constructed for all countries from consumption expenditure series in current and constant monetary units. Although price indices for energy are available, we use the energy price aggregator function described earlier to generate an endogenous price of energy that is a function of fuel prices.[14] Data were obtained from the OECD's *National Accounts,* the U.N.'s *National Accounts,* the EEC's *National Accounts* publications, and the national statistical yearbooks of individual countries. These price indices describe the movement of real prices over time; relative prices of the various categories of consumption in each country are determined from the purchasing power parity indices.

14. The energy price aggregator requires data on fuel prices. For some countries our data on fuel prices does not go back as far as our consumption expenditure data. In these cases the estimated energy price aggregator was regressed against the implicit energy price index, so that data for the index could be used to extend the aggregator backwards.

*Fuel Expenditures* Data were collected for total residential consumption expenditures on petroleum products (largely light fuel oil), natural gas, coal, and electricity. The data through 1970 were obtained from the national accounts publications put out by the statistical office of the EEC, or from the national statistical yearbooks of individual countries. In a few cases for data before 1971, and for all data covering 1971 to 1974, figures were obtained by multiplying the retail price of the fuel by the physical quantity of the fuel consumed, with physical quantity data obtained from the OECD's *Energy Statistics* tape. The data are measured in current local currency units and are converted into U.S. dollars using the purchasing power parity index for residential energy use.

*Fuel Prices* For each fuel, the data represent countrywide averages of the average retail price. In the cases of natural gas and electricity, the problem of differing rate structures in different countries was treated by choosing a price level equivalent to the average price facing an average size household.[15] For those cases where the price of natural gas differs from that of manufactured gas, an average of the prices weighted by the relative amounts consumed was calculated. Data were obtained from the Statistical Office of the EEC's *Energy Statistics, Studien und Erhabungen,* publications and documents of the International Energy Agency, the *Petroleum Times,* and from the national statistical yearbooks of individual countries. All prices were originally measured in local currency units per Tcal of energy but were converted to 1970 U.S. dollars per Tcal.

15. In some studies of fuel demand, for example, that of Halvorsen (76), the marginal price of electricity has been used as a measure of price. This could be a correct approach if expenditures on electricity are a sufficiently small fraction of consumers' budgets so that consumers indeed respond only to the marginal price. However, as Taylor (154) argues, it is more likely that consumers respond to the entire rate structure, so that the use of a marginal price alone is inappropriate, and a correct procedure would be to use the average price at a normalized and constant rate of consumption, or to incorporate both average and marginal prices. It is also possible that the perceived price to consumers is the average cost of a kilowatthour. We do not attempt to resolve this issue here. In this study we use only the average prices of natural gas and electricity since those are the only prices for which data are available.

*Fuel Quantities* The logit models of interfuel substitution as well as the logarithmic models used to model the demands for fuels in some of the developing countries require physical quantity data for fuels. Our data for fuel quantities are implicitly derived from data on fuel expenditures and fuel prices. The unit of measurement is Tcals.

*Other Variables* Data were also collected for net disposable income, population, and temperature. The income data represent total net disposable income of all households, although for some countries only total private income data (personal income plus income going to nonprofit institutions) were available. (These data are put into per capita terms for estimation purposes.) Income data were obtained from the OECD's *National Accounts,* and have been converted to 1970 U.S. dollars. Data for the total population of each country came from the U.N.'s *Demographic Yearbook* and are measured in millions of people. Finally, our temperature data represent the average temperature over the five winter months (November through March) averaged over the principle city or cities of each country. The source is the U.S. Weather Bureau's *Monthly Climatic Data for the World,* and the units of measurement are degrees Fahrenheit.

In table 3.3 we show the range of the data that have been collected for all of the variables. This range does not necessarily represent the time bounds used in model estimation, however. In order to have roughly overlapping bounds, only a subset of the data is used for estimation work.

Our static translog models of consumption expenditures require consumption expenditure shares and price indices, and the models of fuel expenditures require data on fuel shares and fuel prices. It is useful to examine some of the share and price data before turning to the estimation results. Data for 1962 and 1972 are shown in table 3.4. The price index for energy is not shown because it is computed from the translog price aggregator using the fuel prices.

Observe from table 3.4 that the shares of energy in consumption expenditures remained fairly constant over time for most countries, but varied by as much as 100 percent across countries (from 0.03 in Canada and Italy to 0.06 in Belgium and the Netherlands). This same pattern is true for the clothing, durables, and transportation shares; only the food and "other" shares change significantly over time, decreasing and increasing respectively in every country. Since most of the variation in shares is across countries, we would expect our models of consumption expenditures to capture long-run elasticities. Fuel shares, on the other hand, show considerable variation both across time and across countries, and it is therefore more difficult to know a priori whether our estimated partial fuel-price elasticities will be short or long term. One way to help determine this is to estimate a model using pooled annual data, and then repeat the estimation using only data at three- or four-year intervals. If the resulting estimates do not differ much, we can be more certain that the estimated elasticities pertain to the long run.

**The Industrial Sector**

Our models of energy demand in the industrial sector explain the breakdown of production costs into expenditures on capital, labor, and energy, and the breakdown of energy expenditures into expenditures on individual fuels. The estimation of these models requires data for capital, labor, and energy price indices and expenditure shares of manufacturing output and for the prices and quantities of petroleum, natural gas, coal, and electricity used in the industrial sector.

Ten countries are included in our sample: Canada, France, Italy, Japan, the Netherlands, Norway, Sweden, the U.K., the U.S., and West Germany. The data collected for these countries are described briefly as follows:

*Expenditures on Labor* Expenditures on labor include wages and salaries plus supplements paid to the manufacturing sector. For some countries, in particular Canada, Italy, the Netherlands, Norway, and West Germany, the data were available from the U.N.'s *Growth of World Industry* (*GWI*). For other countries, where the U.N. publication lacked data on supplements for all years, it was necessary to extend supplements by using the national percentages of supplements indicated by *GWI*, the U.N.'s *National Accounts* or the International Labor Organization's (ILO) *Statistical Yearbook*. For Sweden and the U.K. it was necessary to determine what percentage of total national compensation went to manufacturing, using data from the U.N.'s *National Accounts* and the ILO's *Statistical Yearbook*. Finally, for the U.S., Japan, and France, national statistical yearbooks were used. Data is in local currency units and is converted to 1970 U.S. dollars using the purchasing power parity numbers for gross domestic product.

*Price of Labor* The price of labor was determined implicitly by dividing labor expenditures by total man-hours of employees. Man-hours of employees was calculated for 1967 by multiplying man-hours of operatives by the ratio of numbers of employees to number of operatives. For Canada, Italy, Japan, Norway, Sweden, the U.S., and West Germany, data is from the U.N.'s *Growth of World Industry*. Where *GWI* did not have the information, man-hours were calculated from U.N. data on number of employees and ILO data on average working hours. Then for every country except Norway, a wage index (1967 = 100) from the U.S. Bureau of Labor Statistics, which includes wages and supplements, was used to convert our price/hour for 1967 to a time series, 1955 to 1974. The time series for Norway was available directly from *Growth of World Industry*. (Note that the resulting index is not quality adjusted.)

*Price of Capital Services* We compute a capital service price index separately for nonresidential structures ($P_{NR}$) and producers durables ($P_D$), and aggregate these two series into a final price of capital services using a Divisia index, where the investment shares of nonresidential structures and durables serve as the Divisia weights. The computation of the price of capital services of each component is based on Christensen and Jorgenson (39), that is, we assume that the investment price of an asset $q$ is equal to the present value of its future services evaluated at the service price $P$ (which is the price we wish to ascertain).[16] We also assume that the service from an asset declines geometrically over time. Then disregarding taxes, the asset price is related to the service price by

$$q_t = \sum_{j=t}^{\infty}\left[(1-d)^{j-t}P_{j+1}\prod_{s=t+1}^{j+1}\frac{1}{1+r_s}\right], \tag{3.12}$$

where $d$ is the depreciation rate and $r$ is the appropriate interest rate. From this we can obtain the equations that relate the price

16. See also Hall and Jorgenson (74) and Coen (44).

index for each type of capital service to the corresponding asset price index:

$$P_{NR}(t) = R(t)q_{NR}(t - 1) + d_{NR}q_{NR}(t) - [q_{NR}(t) - q_{NR}(t - 1)] \quad (3.13)$$

$$P_D(t) = R(t)q_D(t - 1) + d_Dq_D(t) - [q_D(t) - q_D(t - 1)] \quad (3.14)$$

Here $R$ is a long-term government bond interest rate (source: *International Finance Statistics* of the IMF), and $q_{NR}$ and $q_D$ are the asset price indices for nonresidential structures and durables.[17]

For some countries (Canada, France, Italy, the Netherlands, the U.K., and the U.S.) asset price indices and depreciation rates were obtained from the recent work of Christensen et al. (32, 33, 34, 35, 36, 37). For the remaining countries it was necessary to compute implicit asset price indices from gross fixed capital formation in current and constant units using national statistical yearbooks, the U.N.'s or OECD's *National Accounts.* Remaining depreciation rates were obtained from life of capital figures in Denison (50), or else implicit rates from OECD's *National Accounts* were used. Asset price indices were deflated and then converted into indices relative to the U.S. using the appropriate purchasing power parity indices. Data on the investment shares (gross fixed capital formation for producer durables and nonresidential structures) used to compute the Divisia index were obtained from national statistical yearbooks, or U.N.'s or OECD's *National Accounts.* Note that this method of computing the price of capital does not take into account differences in corporate tax structures across countries; we simply did not have access to the data needed to take taxes into account. This means, however, that our price index for capital services must be viewed as approximate.

*Expenditures on Capital Services* Expenditures on capital services were determined by subtracting labor expenditures from value added. Data on value added at factor cost was obtained from either the U.S.'s *Growth of World Industry* or *Annual Yearbook.* Value added for France and West Germany was available only at producer costs, and value-added tax data obtained from the EEC's *Tax*

17. For West Germany the discount rate was used as the interest rate since the government bond yield was not available.

*Yearbook* was used to arrive at a factor cost figure. All of this data is measured in local currency units, was deflated using the local GDP price deflator, and converted to U.S. dollars using the purchasing power parity for GDP. Note that this does not include depreciation. Since the concept of depreciation varies between countries and comparable data is not available, the gross figures are used.

*Fuel Quantities* Quantities of fuels used in the industrial sector (excluding energy conversion) are all obtained from OECD energy publications. Two different publications were used, *Energy Balances of OECD Countries: 1960–1974* (Paris, 1976) and *Energy Statistics of OECD Countries*. The 1976 publication is used for data covering the period 1960 to 1974, since it contains the most recent and revised data and clearly excludes chemical feedstocks. These data series are related to those in the earlier OECD publications via simple linear regressions, which, together with the earlier data, are then used to extrapolate our 1960 to 1974 series back to 1955. The U.S. was treated differently from other countries in that the 1976 publication showed a large amount of "crude and natural gas liquids" consumed by industry. Investigations into other publications and consultations with the Paris office of the OECD and the International Studies Division of the Federal Energy Administration have led us to conclude that this category probably erroneously contains some petroleum products used for petrochemical feedstocks, nonpetroleum hydrocarbons, and other refinery gas. To keep our accounting consistent with other countries, this category was not included in our petroleum total. The unit of measurement for all quantity data is Tcals.

*Fuel Prices* Industrial price of heavy fuel oil, natural gas, coal, and electricity were obtained from EEC publications and the OECD's statistical office. These data are measured in local currency units, and are converted to U.S. dollars using the appropriate purchasing power parities. Final units are 1970 U.S. dollars per Tcal.

*Purchasing Power Parities* Purchasing power parities for gross domestic product, producers durables, and nonresidential structures were obtained from Gilbert and Kravis (64), Gilbert et al. (65), and

Kravis et al. (108), and are all bilateral indices with the U.S. as the base country.

Our translog models of industrial energy demand require cost shares and price indices for capital, labor, and energy, expenditure shares and prices for the four fuels, and the value of output. The available range of our data is shown in table 3.5. (Note that for France and the U.K., factor-share data is not available for some of the early years.) It is useful to examine some of the share and price data before turning to the estimation results. Data for 1962 and 1970 are shown in table 3.6. (Note that the energy price index is computed from a translog fuel-share model; the estimation of this model is discussed in chapter 5.) We see from this table that there is considerable variation in fuel expenditure shares across countries, and through time in any one country. Fuel prices also vary considerably across countries, and have generally decreased over time. Factor shares and prices show much more variation across countries than across time, so that our capital, labor, and energy elasticity estimates should probably be viewed as long term.

**The Transportation Sector**

Our models of energy demand in the transportation sector include a simultaneous-equation model of the demand for motor gasoline, as well as simple single-equation, linear-logarithmic models of the demands for diesel fuel, aviation gasoline, and jet fuel. The single-equation models require data only for the quantities and prices of each fuel, and an economic growth variable such as gross domestic product. The model of gasoline demand, on the other hand, explains the stock of automobiles, the average gasoline-burning efficiency of the stock, and the average usage of this stock, and therefore requires more extensive data covering not only quantities and prices for gasoline but also data on additions to and retirements from the stock of cars, the average traffic volume of cars, average efficiency, and a price index for cars.

Some of the data for the model of gasoline demand are available directly, and some of them are constructed from other available data series. Data are available for new registrations of passenger cars and for the stock of cars, but not for retirements from the stock, so that our depreciation rate series must be constructed implicitly. Similarly, data is available for the total traffic volume of passenger cars in each country of our sample, and for the total consumption of gasoline, so that our average fuel efficiency variable is likewise constructed as an implicit series.

Our model of motor gasoline demand is estimated using data covering the following eleven countries: Belgium, Canada, France, Italy, the Netherlands, Norway, Sweden, Switzerland, the U.K., the U.S., and West Germany. Single-equation models for other fuels used in the transportation sector are estimated using data for these countries, as well as data for the following eleven countries: Austria, Australia, Denmark, Finland, Greece, Turkey, Spain, Japan, Argentina, Brazil, and Mexico. Finally, single-equation models of motor gasoline demand are also estimated for this latter group of countries. The data are described briefly as follows:

*Consumption of Motor Gasoline* Data on the annual consumption of motor gasoline were obtained from the OECD's *Energy Statistics: 1950–1973,* and converted to units of Tcals. For a detailed discussion of the characteristics of this data, see chapter 5 of the data base in "User's Guide" by Carson (28).

*Price of Gasoline* Data for the price of regular motor gasoline were obtained from the international energy statistics compiled by the Federal Energy Administration. The original data were in nominal local currency units per gallon, and were converted to constant local currency units per Tcal using the Gross Domestic Product deflator of each country. Finally, the data were converted into 1970 U.S. dollars per Tcal using purchasing power parity indices for aggregate consumption.

*New Registrations of Passenger Cars* Data were collected for the number of passenger cars registered for the first time during each year. Sources of this data varied from country to country, and include *World Road Statistics, The Motor Industry in Great Britain, World Motor Vehicle Data,* and *Highway Statistics.* The data are described in more detail in chapter 4 of the "User's Guide" (28).

*Stock of Passenger Cars* Our data on the stock of passenger cars is the number of cars registered in a given country as of December

31 in each year. The sources for this data are the same as those for the new registrations category.

*Automobile Depreciation Rate* Our accounting identity for the stock of cars requires an average depreciation rate, or the fraction of the stock of cars removed from the road each year. We derive implicit data series for the depreciation rate of the stock of passenger cars using our data on stock changes and new registrations. The equation is

$$DEPRATE_t = \frac{STK_{t-1} - STK_t + NR_t}{STK_{t-1}}, \tag{3.15}$$

where *DEPRATE* is the depreciation rate, *STK* is the stock of passenger cars, and *NR* is new registrations of passenger cars.

*Traffic Volume per Car* Data were collected for the total traffic volume of passenger cars in each country and in each year, measured in millions of vehicle-kilometers. The source of this data is *World Road Statistics,* and the data are based on actual traffic counts in each country. Our series for traffic volume per car (*TVPC*) were constructed as the ratio of total traffic volume to the stock of cars.

*Average Fuel Efficiency* This variable is calculated implicitly as the ratio of total traffic volume to the total consumption of gasoline. Its units are therefore millions of vehicle-kilometers per Tcal. (This can be converted to miles per U.S. gallon by multiplying by 19.56.)

*Automobile Price Index* The price index for automobiles is constructed in two steps. First, we take the ratio of expenditures on personal transportation equipment in current purchasers' value (local currency units) to the same expenditures in purchasers' value of a base year (1970), thereby obtaining a price index in current local currency units.[18] Next, this ratio is divided by the GDP price deflator for each country, and the resulting real price index is scaled

18. Taking the 1975 price relative to 1970 as an example, we have the ratio $P^{75}Q^{75}/P^{70}Q^{75} = P^{75}/P^{70}$, which is just the nominal price index.

so that it is equal to 100 in 1970. The source of data for expenditures on personal transportation equipment was the U.N.'s *Yearbook of National Accounts Statistics*.

*Population* The data for population came from the *Demographic Yearbook of the United States*, and were measured in millions of persons.

*Gross Domestic Product* Gross Domestic Product series in real local currency units were obtained from the OECD's *National Accounts Statistics*, and were converted into 1970 U.S. dollars using purchasing power parity indices for GDP. We also obtained an implicit Gross Domestic Product deflator from the nominal and real GDP series.

*Quantities and Prices of Other Petroleum Products* Quantity and price data were also collected for diesel fuel, aviation gas, and jet fuel. For the OECD countries, the quantity data were obtained from the OECD's *Energy Statistics 1950–1973*, and for other countries they were obtained from the U.N.'s *World Energy Supplies 1950–1974*. The data on prices were more limited, and came from a number of different sources. These data and sources are described in detail in the data base in "User's Guide" (28).

**Table 3.1**
Base-year (1970) purchasing power parities for each category of residential consumption

| Country | Total consumption | Apparel | Food | Durables[a] | Energy[b] |
|---|---|---|---|---|---|
| Belgium | 43.50 | 38.60 | 48.18 | 43.44 | 43.50 |
| Canada | 1.08 | 0.88 | 1.13 | 0.85 | 1.08 |
| France | 4.64 | 4.23 | 5.49 | 4.00 | 5.13 |
| Italy | 493.00 | 385.03 | 616.65 | 423.47 | 465.80 |
| Netherlands | 2.75 | 2.34 | 3.03 | 2.30 | 1.47 |
| Norway | 6.36 | 5.06 | 8.01 | 4.34 | 4.30 |
| U.K. | 0.31 | 0.24 | 0.33 | 0.22 | 0.34 |
| U.S. | 1.00 | 1.00 | 1.00 | 1.00 | 1.00 |
| West Germany | 3.32 | 2.40 | 3.88 | 2.61 | 2.84 |

[a] For Belgium, Canada, Netherlands, and Norway, no "durables" PPP exists. A PPP for "other household" was used in going from these countries to Germany, and the "durables" PPP was used in bridging from Germany to the U.S.
[b] For Belgium and the Netherlands, the PPP relative to Germany refers to "electricity, gas, and water."
[c] Other set equal to total consumption.

| Transportation and communication | Other[c] |
|---|---|
| 46.03 | 43.50 |
| 1.08 | 1.08 |
| 5.99 | 4.64 |
| 558.85 | 493.00 |
| 3.32 | 2.75 |
| 6.12 | 6.36 |
| 0.31 | 0.31 |
| 1.00 | 1.00 |
| 3.68 | 3.32 |

**Table 3.2**
Alternative engineering estimates of thermal efficiencies

| Citation of Use of Estimate | Adams and Miovic (3) | | Adams and Griffin (2) | |
|---|---|---|---|---|
| Fuel | Residential | Industrial | Residential | Industrial |
| Gas | 0.65-0.72 | 0.39 | 0.60 | 0.65 |
| Solid fuel | 0.05-0.60 | 0.33 | 0.50 | 0.45 |
| Liquid fuel | 0.65 | 0.40 | 0.65 | 0.59 |
| Electricity | 0.80 | 0.80 | — | 0.80 |

| Nordhaus (123) | | Fuss and Waverman (62) | |
|---|---|---|---|
| Residential | Industrial | Residential | Industrial |
| 0.70 | 0.85 | 0.75 | 0.85 |
| 0.20 | 0.70 | 0.50 | 0.87 |
| 0.60 | 0.80 | 0.65 | 0.87 |
| 0.95 | 0.99 | 1.00 | 1.00 |

**Table 3.3**
Range of data for residential energy demand

| | Belgium | Canada | France | Italy |
|---|---|---|---|---|
| **Consumption expenditures** | | | | |
| Food | 55–74 | 50–74 | 58–74 | 55–74 |
| Clothing | 55–74 | 50–74 | 58–74 | 51–74 |
| Durables | 55–74 | 55–74 | 58–74 | 55–74 |
| Transportation and communication | 55–74 | 50–74 | 58–74 | 51–74 |
| Energy | 60–74 | 61–74 | 60–74 | 60–74 |
| "Other" | 60–74 | 61–74 | 60–74 | 60–74 |
| **Consumption expenditure price series** | | | | |
| Food | 55–74 | 50–74 | 58–74 | 55–74 |
| Clothing | 55–74 | 50–74 | 58–74 | 55–74 |
| Durables | 55–74 | 50–74 | 58–74 | 55–74 |
| Transportation and communication | 55–74 | 53–74 | 58–74 | 55–74 |
| Energy | 60–74 | 61–74 | 60–74 | 60–74 |
| "Other" | 55–74 | 50–74 | 58–74 | 55–74 |
| **Fuel expenditures** | | | | |
| Electricity | 60–74 | 50–74 | 60–74 | 60–74 |
| Liquid | 60–74 | 58–74 | 60–74 | 60–74 |
| Solid | 60–74 | 61–74 | 60–74 | 59–74 |
| Gas | 60–74 | 50–74 | 60–74 | 60–74 |
| **Fuel prices** | | | | |
| Electricity | 55–74 | 58–74 | 58–74 | 60–74 |
| Liquid | 55–74 | 58–74 | 58–74 | 57–74 |
| Solid | 58–74 | 55–74 | 58–74 | 55–74 |
| Gas | 55–74 | 58–74 | 58–74 | 55–74 |
| **Miscellaneous** | | | | |
| Personal disposable income | 55–74 | 50–74 | 50–74 | 51–74 |
| Population | 50–74 | 50–74 | 50–74 | 50–74 |
| Temperature | 55–74 | 54–74 | 56–74 | 55–74 |

| Netherlands | Norway | U.K. | U.S. | West Germany |
|---|---|---|---|---|
| | | | | |
| 55–74 | 55–74 | 55–74 | 50–74 | 50–74 |
| 55–74 | 50–74 | 57–74 | 50–74 | 55–74 |
| 55–74 | 55–74 | 54–74 | 50–74 | 55–74 |
| 53–74 | 53–74 | 55–74 | 50–74 | 55–74 |
| | | | | |
| 50–74 | 50–74 | 57–74 | 50–74 | 59–74 |
| 53–74 | 53–74 | 57–74 | 50–74 | 59–74 |
| | | | | |
| 55–74 | 55–74 | 57–74 | 50–74 | 55–74 |
| 55–74 | 53, 55–74 | 57–74 | 50–74 | 55–74 |
| 55–74 | 53, 55–74 | 57–74 | 50–74 | 55–74 |
| 55–74 | 53, 55–74 | 57–74 | 50–74 | 55–74 |
| | | | | |
| 60–74 | 64–74 | 60–74 | 60–74 | 60–63 |
| | | | | 65–74 |
| 55–74 | 53, 55–74 | 57–74 | 50–74 | 55–74 |
| | | | | |
| 60–74 | 50–74 | 57–74 | 50–74 | 55–74 |
| 60–74 | 50–74 | 57–74 | 50–74 | 56–74 |
| 60–74 | 50–74 | 57–74 | 50–74 | 55–74 |
| 59–74 | 50–74 | 57–74 | 50–74 | 59–74 |
| | | | | |
| 55–74 | 64–74 | 57–74 | 52–74 | 55–74 |
| 55–74 | 62–74 | 60–74 | 56–74 | 56–74 |
| 55–74 | 53, 55–74 | 57–74 | 50–74 | 55–74 |
| 55–74 | 55–74 | 57–74 | 59–74 | 55–74 |
| | | | | |
| 50–74 | 55–74 | 53–74 | 50–74 | 50–74 |
| | | | | |
| 50–74 | 50–74 | 51–74 | 50–74 | 50–74 |
| 55–74 | 55–74 | 55–74 | 55–74 | 55–74 |

**Table 3.4**
Share and price data for the residential sector

| | | Expenditure shares | | | | | | Expenditure price indices[a] | | | | |
|---|---|---|---|---|---|---|---|---|---|---|---|---|
| | | Clothing | Food | Durables | Transportation and communication | Energy | Other | Food | Clothing | Durables | Transportation and communication | Other |
| Belgium | 1962 | 0.10 | 0.32 | 0.11 | 0.09 | 0.06 | 0.32 | 1.09 | 0.82 | 1.11 | 0.92 | 1.03 |
| | 1972 | 0.08 | 0.29 | 0.15 | 0.11 | 0.06 | 0.31 | 1.12 | 0.88 | 1.01 | 1.05 | 1.06 |
| Canada | 1962 | 0.09 | 0.25 | 0.08 | 0.14 | 0.03 | 0.41 | 1.02 | 0.79 | 0.81 | 0.90 | 0.94 |
| | 1972 | 0.09 | 0.23 | 0.09 | 0.15 | 0.03 | 0.42 | 1.03 | 0.83 | 0.83 | 1.04 | 0.97 |
| France | 1962 | 0.10 | 0.38 | 0.09 | 0.09 | 0.04 | 0.30 | 1.13 | 0.79 | 0.86 | 1.23 | 1.19 |
| | 1972 | 0.09 | 0.27 | 0.08 | 0.11 | 0.04 | 0.41 | 1.17 | 0.91 | 0.91 | 1.32 | 1.00 |
| Italy | 1962 | 0.093 | 0.44 | 0.08 | 0.08 | 0.03 | 0.28 | 1.22 | 0.75 | 0.88 | 1.05 | 1.01 |
| | 1972 | 0.10 | 0.32 | 0.06 | 0.11 | 0.031 | 0.38 | 1.26 | 0.78 | 0.93 | 1.18 | 0.99 |
| Netherlands | 1962 | 0.12 | 0.34 | 0.15 | 0.04 | 0.05 | 0.30 | 1.05 | 0.82 | 0.81 | 1.10 | 1.17 |
| | 1972 | 0.10 | 0.27 | 0.11 | 0.08 | 0.06 | 0.38 | 1.17 | 0.83 | 0.88 | 1.21 | 1.01 |
| Norway | 1962 | 0.15 | 0.37 | 0.11 | 0.10 | 0.03 | 0.24 | 1.18 | 0.72 | 0.72 | 0.81 | 0.94 |
| | 1972 | 0.1 | 0.30 | 0.08 | 0.12 | 0.04 | 0.36 | 1.27 | 0.79 | 0.71 | 0.96 | 0.99 |
| U.K. | 1962 | 0.09 | 0.37 | 0.08 | 0.10 | 0.05 | 0.31 | 1.04 | 0.65 | 0.79 | 0.91 | 0.94 |
| | 1972 | 0.08 | 0.31 | 0.08 | 0.14 | 0.05 | 0.34 | 1.1 | 0.75 | 0.73 | 1.04 | 1.01 |
| U.S. | 1962 | 0.08 | 0.26 | 0.14 | 0.03 | 0.04 | 0.45 | 0.99 | 0.97 | 1.14 | 0.95 | 0.94 |
| | 1972 | 0.08 | 0.21 | 0.15 | 0.04 | 0.04 | 0.48 | 1.01 | 0.99 | 1.03 | 1.03 | 1.01 |
| West | 1962 | 0.13 | 0.32 | 0.14 | 0.10 | 0.04 | 0.27 | 1.16 | 0.65 | 0.83 | 1.01 | 1.03 |
| Germany | 1972 | 0.114 | 0.284 | 0.14 | 0.12 | 0.06 | 0.282 | 1.18 | 0.71 | 0.82 | 1.13 | 0.96 |

[a] U.S. 1970 = 1.00.
[b] Units: 1970 U.S. \$/Tcal.

| Fuel shares | | | | Fuel prices[b] | | | |
|---|---|---|---|---|---|---|---|
| Solid fuel | Liquid fuel | Gas | Electricity | Solid fuel | Liquid fuel | Gas | Electricity |
| 0.56 | 0.10 | 0.15 | 0.19 | 11069.70 | 6447.2 | 26241.30 | 86837.30 |
| 0.23 | 0.30 | 0.20 | 0.27 | 11334.70 | 5204.8 | 22485.2 | 60981.50 |
| 0.03 | 0.38 | 0.16 | 0.43 | 5623.65 | 4598.48 | 4589.02 | 19177.60 |
| 0.01 | 0.36 | 0.17 | 0.46 | 6197.83 | 4656.45 | 3452.15 | 16247.50 |
| 0.44 | 0.08 | 0.24 | 0.24 | 13555.70 | 6138.03 | 28675.80 | 70642.0 |
| 0.09 | 0.41 | 0.20 | 0.30 | 11772.40 | 6058.19 | 25380.90 | 55661.70 |
| 0.16 | 0.09 | 0.28 | 0.47 | 8940.68 | 5311.31 | 26918.90 | 76763.9 |
| 0.03 | 0.22 | 0.26 | 0.49 | 9784.66 | 3902.12 | 19798.00 | 46565.10 |
| 0.35 | 0.25 | 0.15 | 0.25 | 10857.10 | 5184.33 | 23915.0 | 61273.1 |
| 0.02 | 0.20 | 0.60 | 0.18 | 11228.20 | 4482.82 | 11803.10 | 32805.20 |
| 0.09 | 0.14 | 0.01 | 0.76 | 6831.32 | 6529.05 | 13293.80 | na |
| 0.07 | 0.20 | 0.01 | 0.72 | 8500.93 | 4822.79 | 19986.80 | 12839.30 |
| 0.39 | 0.07 | 0.18 | 0.36 | 6096.33 | 8084.18 | 19472.60 | 35356.40 |
| 0.18 | 0.06 | 0.28 | 0.48 | 7607.69 | 7489.38 | 12215.0 | 30656.60 |
| 0.01 | 0.32 | 0.26 | 0.41 | 8506.91 | 5527.97 | 3535.53 | 34725.20 |
| 0 | 0.30 | 0.23 | 0.47 | 7907.33 | 6077.64 | 3760.37 | 24909.2 |
| 0.11 | 0.20 | 0.11 | 0.58 | 10063.40 | 5271.72 | 24685.10 | 66068.60 |
| 0.03 | 0.22 | 0.21 | 0.54 | 11116.60 | 3661.70 | 31729.10 | 44369.7 |

**Table 3.5**
Range of data for the industrial sector

| Price | Canada | France | Italy | Japan | Netherlands |
|---|---|---|---|---|---|
| **Fuel price** | | | | | |
| Solid fuel | 55–74 | 55–74 | 55–74 | 55–75 | 55–74 |
| Liquid fuel | 55–75 | 55–74 | 55–75 | 55–74 | 55–74 |
| Gas | 55–74 | 55–74 | 55–74 | 55–75 | 55–75 |
| Electricity | 55–74 | 55–74 | 55–74 | 55–74 | 55–74 |
| **Fuel consumption** | | | | | |
| Solid fuel | 55–74 | 55–74 | 55–74 | 55–74 | 55–74 |
| Liquid fuel | 55–74 | 55–74 | 55–74 | 55–74 | 55–74 |
| Gas | 55–74 | 55–74 | 55–74 | 55–74 | 55–74 |
| Electricity | 55–74 | 55–74 | 55–74 | 55–74 | 55–74 |
| **Expenditure on fuels** | | | | | |
| Solid fuel | 55–74 | 55–74 | 55–74 | 55–74 | 55–74 |
| Liquid fuel | 55–74 | 55–74 | 55–74 | 55–74 | 55–74 |
| Gas | 55–74 | 55–74 | 55–74 | 55–74 | 55–74 |
| Electricity | 55–74 | 55–74 | 55–74 | 55–74 | 55–74 |
| **Other** | | | | | |
| Capital price | 55–73 | 55–73 | 55–73 | 55–73 | 55–73 |
| Labor expenditures | 58–74 | 60–74 | 61–73 | 58–74 | 58–74 |
| Labor price | 60–74 | 60–73 | 60–73 | 60–73 | 60–73 |
| Value added | 58–73 | 63–73 | 58–73 | 58–73 | 58–72 |

| Norway | Sweden | U.K. | U.S. | West Germany |
|---|---|---|---|---|
| 55-74 | 55-74 | 55-74 | 54-7, 59-74 | 55-74 |
| 55-75 | 55-74 | 55-74 | 54-7, 59-74 | 55-74 |
| 55-74 | 55-75 | 55-74 | 54-7, 59-74 | 55-75 |
| 55-75 | 55-74 | 55-74 | 54-7, 59-74 | 55-74 |
| 55-74 | 55-74 | 55-74 | 55-74 | 55-74 |
| 55-74 | 55-74 | 55-74 | 55-74 | 55-74 |
| 55-74 | 55-74 | 55-74 | 55-74 | 55-74 |
| 55-74 | 55-74 | 55-74 | 55-74 | 55-74 |
| 55-74 | 55-74 | 55-74 | 55-74 | 55-74 |
| 55-74 | 55-74 | 55-74 | 55-74 | 55-74 |
| 55-74 | 55-74 | 55-74 | 55-74 | 55-74 |
| 55-74 | 55-74 | 55-74 | 55-74 | 55-74 |
| 56-73 | 58-73 | 56-73 | 55-73 | 58-73 |
| 58-74 | 60-74 | 60-74 | 58-73 | 60-73 |
| 58-73 | 60-74 | 60-74 | 60-74 | 60-74 |
| 53, 58-73 | 53, 58 60-73 | 58, 63-73 | 58-73 | 58-73 |

**Table 3.6**
Share and price data for the industrial sector

| | | Fuel expenditure shares | | | | Fuel prices (in 1970 U.S. dollars/Tcal) | | | |
|---|---|---|---|---|---|---|---|---|---|
| | | Solid fuel | Liquid fuel | Gas | Electricity | Solid fuel | Liquid fuel | Gas | Electricity |
| Canada | 1962 | 0.14 | 0.23 | 0.07 | 0.57 | 1950.77 | 3693.84 | 1658.82 | 9472.0 |
| | 1970 | 0.09 | 0.26 | 0.12 | 0.52 | 1919.33 | 3187.48 | 1498.22 | 8058.0 |
| France | 1962 | 0.34 | 0.14 | 0.10 | 0.42 | 3986.67 | 3738.77 | 11107.4 | 21816.3 |
| | 1970 | 0.23 | 0.19 | 0.06 | 0.52 | 3953.2 | 1835.1 | 3982.76 | 20218.9 |
| Italy | 1962 | 0.13 | 0.22 | 0.08 | 0.56 | 6123.38 | 4476.98 | 3638.33 | 33163.1 |
| | 1970 | 0.12 | 0.23 | 0.09 | 0.56 | 5846.13 | 2711.22 | 2898.58 | 23147.9 |
| Japan | 1962 | 0.19 | 0.15 | 0.08 | 0.57 | 4317.05 | 4102.77 | 2980.97 | 30592.1 |
| | 1970 | 0.13 | 0.20 | 0.03 | 0.65 | 3065.6 | 2696.5 | 1832.62 | 24639.3 |
| Netherlands | 1962 | 0.14 | 0.18 | 0.001 | 0.67 | 4268.71 | 3281.06 | 8593.59 | 66391.4 |
| | 1970 | 0.06 | 0.10 | 0.15 | 0.69 | 3384.83 | 2194.8 | 2572.89 | 37697.1 |
| Norway | 1962 | 0.09 | 0.19 | 0 | 0.74 | 2533.29 | 2922.16 | 10250.3 | 6006.92 |
| | 1970 | 0.10 | 0.30 | 0 | 0.61 | 3479.78 | 3305.39 | 16817.3 | 5318.52 |
| Sweden | 1962 | 0.21 | 0.14 | 0.01 | 0.66 | 4037.07 | 3364.28 | 10933.4 | 26752.4 |
| | 1970 | 0.15 | 0.24 | 0.01 | 0.60 | 3015.79 | 2826.3 | 6540.04 | 15715.9 |
| U.K. | 1962 | 0.37 | 0.15 | 0.07 | 0.46 | 3543.11 | 3774.79 | 11949.7 | 29415.0 |
| | 1970 | 0.16 | 0.21 | 0.06 | 0.56 | 2512.06 | 2581.53 | 5724.2 | 24269.2 |
| U.S. | 1962 | 0.1 | 0.08 | 0.19 | 0.64 | 1120.63 | 1749.0 | 1376.74 | 14254.8 |
| | 1970 | 0.06 | 0.07 | 0.21 | 0.65 | 1011.43 | 1541.05 | 1111.11 | 11059.3 |
| West Germany | 1962 | 0.31 | 0.11 | 0.003 | 0.58 | 4079.97 | 3360.8 | 4252.76 | 33188.8 |
| | 1970 | 0.15 | 0.15 | 0.05 | 0.64 | 4053.35 | 2415.6 | 3201.81 | 26605.8 |

| Capital | | Labor | | Energy | | |
|---|---|---|---|---|---|---|
| Expenditure share | U.S. 1970 = 1 price | Expenditure share | U.S. 1970 = 1 price | Expenditure share | U.S. 1970 = 1 price | Net value output |
| 0.422 | 0.77 | 0.523 | 0.61 | 0.055 | 1.10 | 166.6 |
| 0.39 | 0.999 | 0.56 | 0.75 | 0.05 | 0.94 | 226.65 |
| na | 1.17 | na | 0.30 | na | 1.75 | na |
| 0.45 | 1.22 | 0.50 | 0.40 | 0.05 | 1.73 | 499.69 |
| 0.47 | 1.39 | 0.45 | 0.33 | 0.08 | 2.9 | 234.96 |
| 0.363 | 1.15 | 0.564 | 0.53 | 0.072 | 2.04 | 333.24 |
| 0.54 | 1.47 | 0.38 | 0.21 | 0.08 | 2.72 | 524.19 |
| 0.57 | 1.49 | 0.37 | 0.41 | 0.06 | 2.21 | 1232.52 |
| 0.37 | 0.78 | 0.56 | 0.36 | 0.07 | 5.5 | 92.23 |
| 0.404 | 0.82 | 0.533 | 0.59 | 0.063 | 3.27 | 134.03 |
| 0.45 | 0.78 | 0.5 | 0.45 | 0.05 | 0.95 | 28.39 |
| 0.41 | 1.07 | 0.52 | 0.62 | 0.064 | 0.93 | 37.64 |
| 0.33 | 1.35 | 0.57 | 0.57 | 0.10 | 2.68 | 78.45 |
| 0.35 | 1.45 | 0.584 | 0.87 | 0.064 | 1.66 | 119.01 |
| na | 1.10 | na | 0.33 | na | 2.16 | na |
| 0.35 | 1.06 | 0.6 | 0.42 | 0.05 | 1.81 | 622.02 |
| 0.434 | 1.10 | 0.533 | 0.86 | 0.03 | 1.29 | 2550.72 |
| 0.46 | 1.00 | 0.51 | 1.00 | 0.03 | 1.00 | 3291.5 |
| 0.62 | 1.26 | 0.33 | 0.33 | 0.05 | 2.91 | 673.61 |
| 0.553 | 1.06 | 0.403 | 0.50 | 0.044 | 2.45 | 921.7 |

# 4 The Residential Demand for Energy

The characteristics of energy demand in the residential sector depend upon the role of energy in the consumption basket (and in particular the willingness of consumers to substitute between energy and other categories of consumption), as well as the substitutability of individual fuels. These aspects of residential energy demand should be captured by the two-stage model developed in chapter 2. Recall that in the first stage of this model consumers choose utility-maximizing expenditure shares for energy and five other consumption categories, while in the second stage energy expenditures are broken down into expenditures on oil, natural gas, coal, and electricity, again under the assumption that these shares are chosen to maximize utility.

In this chapter we present the statistical results from estimating the various versions of our residential demand model, we compare these results to those of other studies, and we discuss the implications of the statistical evidence for the characteristics and likely future behavior of residential energy demand. The most successful results were obtained using the static indirect translog utility function with time-varying preferences for each stage of the model, so that the estimation of the translog function occupies most of this chapter. However, we have also estimated static and dynamic logit models for fuel shares, and these results are also discussed here.

Let us now review our procedure for applying the indirect translog utility function to our two-stage model of residential energy demand. We begin by estimating the translog price aggregator that will be used to obtain an aggregate price index for energy. This involves estimating the share equations corresponding to a homothetic translog cost function (equivalent to a homothetic and stationary indirect utility function with unit total expenditure):

$$S_i = \gamma_i + \sum_j \gamma_{ij} \log P_j, \qquad i = 1, \ldots, 4. \tag{4.1}$$

These equations are estimated subject to the parameter restrictions $\Sigma\gamma_i = 1$, $\gamma_{ij} = \gamma_{ji}$, and $\Sigma_i\gamma_{ij} = 0$. As explained in chapter 3, we estimate this aggregator in two different ways, first using four fuel categories for all nine countries, and second "bridging" a three-fuel aggregation (oil, gas, and electricity) for the U.S. and Canada with a four-fuel aggregation for the remaining countries. One of the resulting aggregators is used to generate the aggregate price index for energy in the residential sector of each country. A relative price index is then calculated for every country in every year.

Next the expenditure-share equations

$$S_j = \frac{\alpha_j + \sum_i \beta_{ji} \log (P_i/M) + \beta_{jt} \cdot t}{\alpha_M + \sum_i \beta_{Mi} \log (P_i/M) + \beta_{Mt} \cdot t}, \tag{4.2}$$

are estimated for the six categories of consumption expenditures. The estimated aggregated price index for energy is used as an instrumental variable for the price of energy in the estimation of these equations. The share equations are estimated in stages, with additional parameter restrictions imposed and tested at each stage. We begin by testing the symmetry restrictions $\beta_{ij} = \beta_{ji}$ implied by utility maximization. We next test for explicit homotheticity, namely, the restrictions that the $\beta_{Mi} = 0$ and $\beta_{Mt} = 0$, for stationarity, such that the $\beta_{jt} = 0$, and the restriction that energy is explicitly groupwise separable from the other categories of consumption, such that the parameters $\beta_{Ej}, j = 1, \ldots, 6$, are all zero. In addition, we test the restrictions of additivity and the restrictions of regional homogeneity, that is, that various parameters are the same across countries. After estimating the share equations, the parameters are used to obtain elasticities of demand for energy and the other consumption categories.

We next apply the share equations (4.2) to the breakdown of energy expenditures into expenditures on individual fuels. In estimating these equations, the restrictions of homotheticity are imposed a priori, as required for a two-stage model of utility maximization. However, conditional on homotheticity, we test restrictions of stationarity and additivity, as well as cross-country homogeneity.

Nine countries are included in our sample: Belgium, Canada, France, Italy, the Netherlands, Norway, the U.K., the U.S., and West Germany. All of these share equations are estimated using data that span the period 1960 to 1974.

In estimating the expenditure share equations, certain decisions must be made regarding the use of dummy variables and the choice of time bounds. First, recall that regional dummy variables can be used to allow the first-order parameters $\alpha_j$ of the share equations to vary across countries, or to allow the second-order parameters $\beta_{ji}$ to vary across countries. The latter alternative involves a considerable reduction in degrees of freedom, so that resulting elasticities have large standard errors. We therefore report here only the use of regional dummy variables for the first-order parameters.

Second, although our data span the period 1960 to 1974, there is a question as to whether the 1974 data should be considered to have come from the same population as the earlier data, that is, whether the 1974 data point corresponds to the same long-run indirect utility function. We therefore re-estimate some of the models excluding the 1974 data, and by examining the resulting sensitivity of the estimates, determine whether prices and expenditure shares moved off the long-run function in 1974.

Finally, we estimate our models using data at four-year intervals and compare the results with the corresponding ones obtained using annual data. If the resulting estimates are nearly the same, we can conclude that most of the explanation in the data is cross-sectional, so that we are more likely to have obtained long-run estimates of the elasticities.

In estimating the share equations, we ignore autocorrelation of the error terms within equations but account for error correlations across equations. In particular, we use iterative nonlinear Zellner estimation (under the assumption of no heteroscedasticity or autocorrelation within equations), which is equivalent to full-information maximum-likelihood estimation.

### 4.1 Estimation of the Energy Price Aggregator

We begin with the estimation of the translog energy price aggregator. Because very little coal is consumed in the residential sectors of the U.S. and Canada, there is a question as to whether a three-fuel aggregator should be estimated for Canada and the U.S. and then bridged to a four-fuel aggregator for Europe, or a four-fuel aggregator should be estimated for all nine countries. We estimate an energy price index using both of these methods.

First, a translog price aggregator is estimated assuming a choice of four fuels in all nine countries. This version of the aggregator is in turn estimated, first, by restricting all of the parameters to be the same across all countries and, second, including regional intercept dummy variables to allow the parameters $\gamma_i$ to vary across countries. The resulting estimated parameters are shown in columns 1 and 2 respectively in table 4.1. The chi-square statistic is used to test the restrictive hypothesis of regional homogeneity, that is, the $\gamma_i$ parameters constant across countries. The test statistic is 336.0, which, with 24 degrees of freedom, is significant at the 1 percent level, leading us to reject regional homogeneity and include the intercept dummy variables in the estimation of the price aggregator.

Next a price index is estimated using the bridging method. First, a four-fuel aggregator is estimated for the seven European countries (including country dummy variables), and the parameter $\gamma_0$ is chosen so that the price of energy is equal to 1 in Belgium in 1970. The resulting parameter estimates are shown in column 3 of table 4.1. Next, the four-fuel model estimated previously for all nine countries is used to compute the price of energy in Belgium in 1970 relative to that in the U.S.; this ratio is 1.9315. Then the price aggregator is estimated for three fuels for the U.S. and Canada, and the parameter $\gamma_0$ for this two-country aggregator is chosen so that the price of energy is equal to 1 in the U.S. in 1970. The estimation results are shown in column 4 of table 4.1. Finally, using the Belgium-to-U.S. ratio of 1.9315, the four-fuel price indices calculated for the European countries are converted to a U.S. 1970 = 1 base.

The resulting energy price indices computed under the two methods are shown in table 4.2 for all nine countries. Note that for all countries except the U.S. the two indices are quite close to each other. For the U.S. the bridging method yields an unrealistically low price of energy in the early years. This is due in large part to a heavy positive weight attached to natural gas, based on a coefficient ($\gamma_{33}$) that is not statistically significant. We therefore use the four-fuel price index for all countries in estimating the consumption expenditure-share equations.

Because of its importance in the estimation of the aggregate consumption model it is useful to examine the energy price index in more detail. In figure 4.1 the index is plotted over time for each of five countries, France, the Netherlands, the U.K., the U.S., and West Germany. Observe from this figure and from table 4.2 that the variation in the price index across countries is considerable, with the index in some countries about two to four times as large as in others. For most countries, however, this index does not vary very much across time, at least until 1973. (The annual percent change in the index is shown for three countries, the U.K., the

U.S., and West Germany, in figure 4.2.) This means that most of the explanation provided by this data will come from intercountry differences, which reinforces our expectation that we will be estimating long-run elasticities.

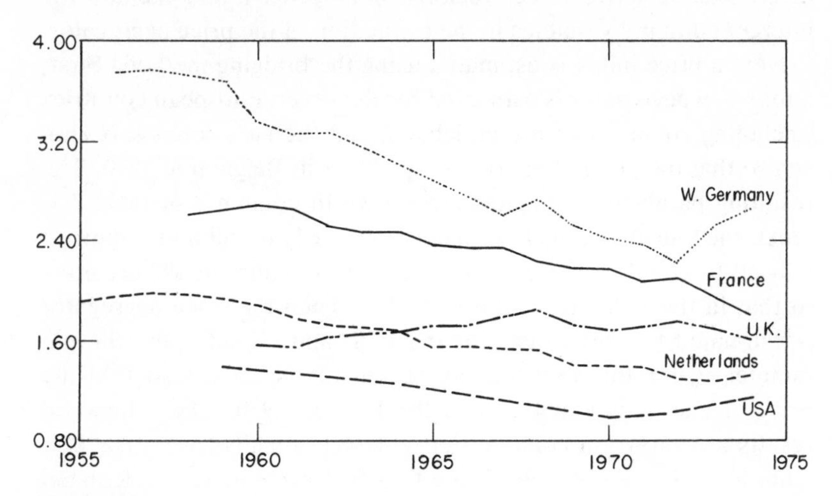


**Figure 4.1**
Energy price index (U.S. = 1 in 1970)

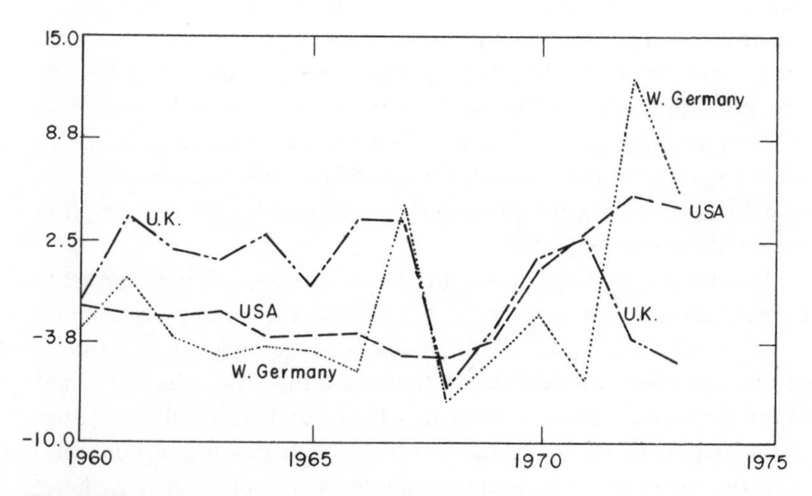


**Figure 4.2**
Annual percent change in energy price index

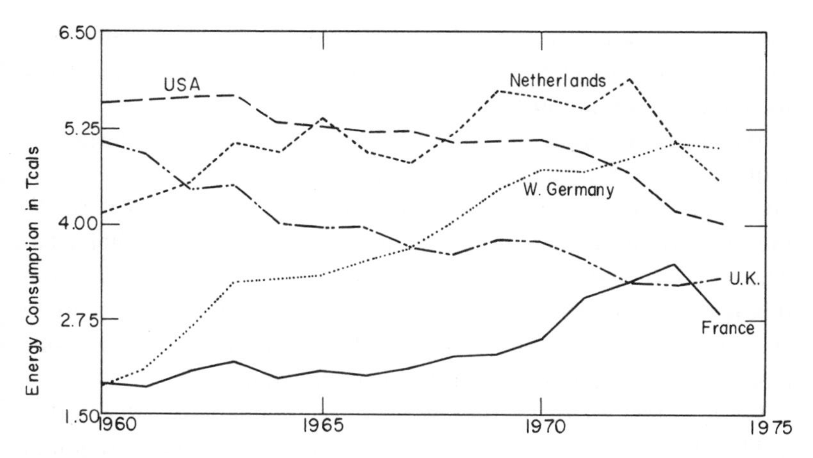


**Figure 4.3**
Residential energy consumption per million dollars of total consumption expenditures

The patterns of residential energy consumption across countries are similar to those of energy prices. In figure 4.3 total residential energy consumption per million 1970 dollars of private consumption expenditures is plotted over time for the same five countries as in figure 4.1. Observe that once again there is considerable intercountry variation in energy use per dollar of consumption expenditures. These intercountry differences in prices and quantities should permit us to obtain reasonably low-variance estimates of the long-run demand elasticities for energy and the other categories of consumption.

## 4.2 The Translog Model of Consumption Expenditures

All of the models discussed here are based on estimates obtained by pooling all nine countries together. Alternative poolings were tested, including separating the U.S. and Canada from the European countries. However, there seemed to be no clear evidence of regional differences in the second-order parameters that would lead us to favor an alternative pooling, and standard errors of the parameter estimates were higher when fewer countries were pooled.

Restrictions on the structure of demand are tested by using the chi-square statistic of equation (2.7) to compare an estimated model that contains a particular restriction to one without the restriction. Parameter estimates for several, but not all of the versions of the model are given in table 4.3.

We begin by testing the basic assumption of utility maximization. Utility maximization implies the symmetry restrictions $\beta_{ij} = \beta_{ji}$, and we test these by estimating and comparing a model for which the restrictions are imposed with one in which they are not imposed. The resulting test statistic is 11.0, and with 10 degrees of freedom this is below the critical 10 percent level, which would allow us to accept these restrictions. We therefore test further restrictions based on the assumption of utility maximization.[1]

We next test the restriction of explicit homotheticity. (Recall that unlike the fuel-share equations, where homotheticity is required because of the two-stage nature of the model, there are no a priori grounds for imposing homotheticity on the consumption expenditure-share equations.) This involves the six parameter restrictions $\beta_{Mj} = 0$, and we also include the homogeneity restriction $\beta_{Mt} = 0$. Parameter estimates for the unrestricted and restricted versions of the model are shown in columns 1 and 2, respectively, of table 4.3.[2] The test statistic is 9.15, and with 7 degrees of freedom this is below the critical 10 percent level, so that these restrictions can be accepted. We therefore test further restrictions conditional on the assumption of homotheticity.

We now test the restriction of stationarity, that is, that the $\beta_{jt} = 0$. A stationary version of the model is estimated, and the resulting parameters are shown in column 3 of table 4.3. The test statistic that compares the models of columns 2 and 3 is 134.8, which, with

1. In applications of the direct and indirect translog utility functions to consumption behavior in the U.S., Christensen, Jorgenson and Lau (42) and Jorgenson and Lau (101) found that the assumption of utility maximization could be rejected. However, they went on to estimate models based on this assumption. This is in fact reasonable, given the nature of the chi-square test. It is a weak test, so that if the test statistic is above a critical level, one could, but need not, reject a particular set of restrictions.
2. Because of the severe nonlinearity of the share equations, considerable computational expense is involved in estimating the nonhomothetic model. We therefore limited the estimation procedure to 150 iterations. This gave estimates that appeared close to those that would have resulted had the procedure been allowed to run to full convergence.

5 degrees of freedom, is well above the critical 1 percent level. We therefore reject stationarity.

We next test for regional homogeneity, that is, that the first-order parameters $\alpha_j$ are the same across all countries. We estimate a homothetic nonstationary model for which the country dummy variables are excluded (the parameter values are not reported here) and compare that to the model in column 2. The resulting test statistic is 1111.5, which is well above the critical 1 percent level, so that we reject regional homogeneity and retain the country dummy variables.[3]

The model in column 4 of table 4.3 has the restriction that energy is explicitly groupwise separable from the other categories of consumption, that is, that the parameters $\beta_{Ej}$, $j = 1, \ldots, 6$, are all zero. We use the chi-square test to compare this model with that of column 2; the test statistic is 14.8, and with 5 degrees of freedom this is above the critical 2.5 percent level, but below the 1 percent level. We therefore reject the restriction of energy separability but recognize that, since several of the $\beta_{Ej}$ parameters are statistically insignificant, the own-price elasticity of total energy consumption will be close to $-1$.

Finally, the model in column 5 is restricted to be explicitly additive, and, since it is also homothetic, this means that all of the $\beta_{ij}$ parameters are zero. Note that additivity implies therefore that all of the own-price elasticities are $-1$ and all of the cross-price elasticities are 0. The test statistic for this model, compared to the unrestricted model in column 2, is 77.1. With a total of 30 degrees of freedom, this is above the critical 1 percent level of 50.1, so that additivity can be rejected.

The results of these tests indicate that a nonstationary model based on a nonadditive indirect utility function with first-order coefficients that vary across countries is needed to estimate price elasticities of consumption expenditures. We therefore retain the model of column 2 as our preferred model. Note that $R^2$ for each of the equations in this model is high, but a good deal of this explanation

3. Although we rejected stationarity, we perform a second test for regional homogeneity by comparing two stationary models, one with and one without the dummy variables. The test statistic is 1082.8, again indicating rejection of the restriction.

can be attributed to the country intercept terms.[4] On the other hand, 25 of the 36 second-order coefficients are significant at the 5 percent level.

As we mentioned earlier, it would also be desirable to estimate our preferred model using data at three- or four-year intervals, as a means of verifying that the implied elasticities indeed pertain to the long run. However, because of the large number of parameters involved in the consumption model, estimation with only one-third of the data is not possible given the reduction in degrees of freedom.[5] It would still be useful to re-estimate the model eliminating some of the data in order to test the stability of the coefficients and implied elasticities to changes in the data set. The model in column 6 of table 4.3 is the same as that in column 2, except that it is estimated using data over the period 1962 to 1973 instead of 1960 to 1974. Comparing columns 2 and 6, we see that most of the estimated parameters are the same. We found that elasticities computed for the two models are also very close to each other. This is encouraging, and indicates that the model is fairly robust and that the full data set, including the 1974 data, can be used for estimation.

The own- and cross-price elasticities corresponding to our preferred consumption model are shown in table 4.4. Since these elasticities depend on the particular prices and shares, they are calculated for each of two years, 1965 and 1973. Income elasticities of demand are not shown; they are all equal to 1.0 since the model is homothetic.

These elasticity estimates are accompanied by standard errors (shown in parentheses below each estimate). These standard errors are in fact approximations to the true standard errors, which cannot be calculated exactly because the elasticities are nonlinear functions of the estimated parameters (since the expenditure shares are themselves functions of the parameters). Since all of the elasticities reported here pertain to homothetic models (for which the parameters $\beta_{Mi}$ and $\beta_{Mt}$ are zero), we can obtain approximate estimates of

4. When the country intercept terms are omitted $R^2$ for these equations falls to 0.453, 0.321, 0.594, 0.334, and 0.083, respectively.

5. On the other hand, because most of the variation in the consumption-share data is cross-sectional rather than time wise, we can be fairly sure that the elasticities are long run.

the standard errors by assuming that the shares $S_i$ are constant at their estimated values in each particular year. Under this assumption we have, asymptotically,

$$\begin{aligned} \mathrm{Var}(\hat{\eta}_{ii}) &= \mathrm{Var}(\hat{\beta}_{ii})/\hat{S}_i^2 \\ \mathrm{Var}(\hat{\eta}_{ij}) &= \mathrm{Var}(\hat{\beta}_{ij})/\hat{S}_i^2. \end{aligned} \tag{4.3}$$

Note that none of the own-price elasticities for energy are significantly different from $-1$ (as would be implied by energy separability). This is reasonable, given that we presume to estimate long-run elasticities, and, as we will see, is within the bounds of alternative estimates obtained by others. For the other categories of consumption, own-price elasticities vary between $-1$ and about $-1.7$. As expected, the smallest own-price elasticity (about $-1.04$) is for food. This may not reflect a true price response, but rather the result of food prices rising slightly with food shares dropping considerably as consumers spend larger incomes on other goods. Cross-price elasticities are presented only for energy in order to save space. Most of these are near zero except for food and transportation. The large cross-price elasticity for food and energy is reasonable; we expect these goods to be substitutes, particularly if incomes are low. The negative cross-price elasticities for energy and transportation are surprising, but again may represent something other than a true price effect; as energy became cheaper during the 1960s, the infrastructure grew that made expansion of the transportation and communication shares possible.

## 4.3 The Translog Model of Fuel Expenditures

We next estimate the share equations for the breakdown of energy expenditures into expenditures on individual fuels. All of the models that we estimate here are homothetic (so that the elasticities of fuel expenditures with respect to total energy expenditures are all equal to 1), but we test restrictions of regional homogeneity, stationarity, and additivity.

Parameter estimates for several versions of this model are given in table 4.5. The parameter estimates given in the first two columns of that table correspond to models that are regionally homogeneous, that is, all of the parameters are the same across all countries. The first model, however, has the additional restriction of stationarity, so that a comparison of the two models provides a first test of stationarity. The test statistic is 43.6, and with 3 degrees of freedom this is well above the critical 1 percent level, leading us (conditional on regional homogeneity) to reject stationarity.

All of the remaining models are regionally heterogeneous—namely, regional dummy variables multiply the first-order (intercept) parameters of the share equations. The models in columns 3 and 4 of table 4.5 have no additional restrictions, but in column 3 the parameters were estimated using data over the entire range 1960 to 1974, whereas in column 4 the parameters were estimated without the 1974 data. By comparing columns 3 and 4, we can determine whether it is appropriate to include the 1974 data. (Since energy prices rose considerably in that year, and since we are estimating what we believe are long-run elasticities, there is some question as to whether the 1974 data belong to the same sample as the 1960 to 1973 data, that is, whether the 1974 data were generated by the same indirect utility function.) Since the estimated parameters in the two columns are quite close to each other (as are the implied elasticities), we conclude that the 1974 data belongs to the same sample, and estimate our models using the full 1960 to 1974 data set.

Comparing the models of columns 3 and 2 provides a test of the regional homogeneity restrictions. The test statistic is 626.0, which is well above the critical 1 percent level, so that we reject homogeneity and include regional dummy variables for the first-order parameters of our model. Now, given a regionally heterogeneous model, we again test for stationarity. Parameter estimates for the stationary version of the model in column 3 are shown in column 7. The appropriate test statistic is 195.4, which is well above the 1 percent level, so that we again reject stationarity. Finally, we test for additivity. In the model shown in column 9, all of the $\beta_{ij}$ parameters are constrained to be 0. Comparing this model to that in column 3, we obtain a test statistic of 189.3. This is above the critical 1 percent level, so that we reject additivity.

Fuel prices have been lower in the U.S. and Canada than in the European countries, and incomes have been higher, which suggests pooling the U.S. and Canada separately. The results of such a pooling are shown in columns 5 (Europe only) and 6 (U.S. and Canada only). The models in this case are regionally heterogeneous and nonstationary. We can compare the resulting parameter estimates (in particular the parameters of the second-order terms in the indirect translog utility function) to those in column 3. We do find that the $\beta_{ij}$ parameters change considerably and that the parameter values for the U.S. and Canada differ considerably from those for the European countries (so that elasticity estimates will also differ considerably). Furthermore, a slightly larger fraction of the $\beta_{ij}$ parameter estimates are statistically significant when the U.S. and Canada are pooled separately. This would argue strongly for a separate pooling of the U.S. and Canada. On the other hand, in pooling these countries separately, we are relying mostly on timewise variation of prices and shares to obtain parameter estimates, which could explain why the parameter estimates in columns 5 and 6 differ by as much as they do; the estimates for Europe may represent a long-run utility function, while those for the U.S. and Canada represent a shorter-run utility function. With this in mind we report elasticities for both versions of the model—the U.S. and Canada pooled separately (table 4.7), and all nine countries pooled together (table 4.8). Indeed, we see that pooling the U.S. and Canada separately results in much smaller price elasticities for these countries than for the European countries, which would be consistent with the estimation of a shorter-run function.

Finally, we wish to verify that pooling the nine countries together yields elasticity estimates that are indeed long run. To do this we estimate the regionally homogeneous (but nonstationary) version of the model using data at four-year intervals (1962, 1966, 1970, 1974). We use a regionally homogeneous model since there are not enough degrees of freedom to identify the 32 additional parameters corresponding to the country dummy variables in the heterogeneous version. The results are shown in column 8, and we can observe that the parameter estimates are quite close to those in column 2, although with larger standard errors. Own-price elasticities for the models in columns 2 and 8 are shown in table 4.6. These estimates are very close to each other, which is evidence that the use of pooled data has elicited long-run elasticities.

Price elasticities for the model in which the U.S. and Canada are pooled separately are shown in table 4.7. (Remember that these are partial elasticities, that is, they are based on a constant total expenditure on energy.) Note that for liquid fuel and gas the elasticities for the U.S. and Canada are about half the size of those for the other countries. This could be because per capita incomes have been much higher in the U.S. and Canada (so that fuel expenditures for home heating are not viewed by consumers as a discretionary component of their consumption baskets), or because we are simply estimating shorter-run elasticities. On the other hand the own-price elasticities of electricity demand for the U.S. and Canada are larger negative numbers than for the European countries (where in most cases they are insignificantly different from 0 or positive). This could be because of greater discretion in the choice of electricity for home heating in the U.S. and Canada, where electricity tends to be used predominantly to heat vacation homes or homes in warmer climates.

Own- and cross-price elasticities for the version of the model in which all nine countries are pooled are shown in table 4.8. Note that own-price elasticities for solid fuel and liquid fuel are all close to $-1$, and show little statistically significant variation across countries. Own-price elasticities for natural gas are larger ($-1.45$ to $-1.99$) and show more regional variation. The magnitudes of these elasticities are reasonable, particularly given that they are long run. The own-price elasticity estimates for electricity are disturbing, however, in that three of them (Belgium, France, and the Netherlands) are positive. This occurs because of the large negative value that is estimated for $\beta_{44}$. This parameter, when divided by the negative of the small expenditure shares for electricity in these countries, becomes greater than 1. Finally, note that all of the cross-price elasticities associated with electricity are negative (and statistically significant).

Although we have rejected the hypothesis of stationarity, it is useful to examine elasticities for the model in which this restriction was imposed. It may be that the time trend variables were statistically significant in estimates of the share equations not because of a change over time in the indirect utility function but because prices and expenditure shares all monotonically increased or decreased. If this was the case the nonstationary model may lead to underestimates of some of the elasticities. We have therefore calculated own- and cross-price elasticities for the stationary version of the fuel-share model (that is, corresponding to column 7 of table 4.5), and these are shown in table 4.9. Note that the own-price elasticities for solid fuel are much larger than for the nonstationary model, perhaps unreasonably large in later years. Own-price elasticities for liquid fuel and natural gas are about the same as in the nonstationary model, but those for electricity are now all negative, and all larger in magnitude.

Although the stationary fuel-share model yields own-price elasticities for electricity that are consistently negative, we follow the results of the statistical tests described earlier and assume that the underlying indirect utility function is indeed nonstationary. As can be seen from table 3.4, there are no clear trends in the fuel prices or shares, and the solid fuel-price elasticities in the stationary model seem unreasonably large in magnitude. Also, we take as our "preferred" model that for which all nine countries have been pooled together. Although pooling the U.S. and Canada separately leads to parameter estimates and elasticities that are quite different from those for the European countries, most elasticities (other than those for electricity) become smaller in magnitude, so that it is more likely that we have elicited short-run elasticities. Our preferred fuel-share model is therefore that of column 3 in table 4.5, and it is this model that we will refer to next when we compare our results to those of other studies.

Total fuel price elasticities are computed using the estimates of the own-price elasticities of aggregate energy use for our preferred consumption model (these elasticities range from $-1.05$ to $-1.15$). We compute total price elasticities corresponding to our preferred fuel-share model, that is, the nonstationary model estimated by pooling all nine countries. These elasticities are shown in table 4.10. Note that since the estimates of the own-price elasticity of energy do not differ very much from $-1$, the total fuel price elasticities are always within 10 percent of the partial elasticities.

## 4.4 Summary of Results and Comparison with Other Studies

As we have seen, the use of the indirect translog utility function in a two-stage model of residential energy demand provides a means of estimating demand elasticities for the energy aggregate (as well as the other component categories of the consumption basket), and for individual fuels. In addition, by pooling international time-series cross-section data we were able to obtain a sample large enough to yield low-variance estimates of essentially long-run elasticities. A summary of some of the more important elasticity estimates that have been obtained here is given in table 4.11.

The use of the translog form also allowed us to test, rather than impose a priori, other restrictions on the structure of demand. In estimating the model of aggregate consumption shares, we found that the data support acceptance of the symmetry restrictions resulting from the hypothesis of utility maximization. We also found that the indirect utility function is homothetic, regionally heterogeneous, and nonstationary, and that no other restrictions can be maintained. Similarly, in estimating the model of fuel shares, we found the indirect utility function to be regionally heterogeneous and nonstationary (conditional on utility maximization and homotheticity), and that no other restrictions could be supported by the data.

One of our more important results is that the own-price elasticity of aggregate energy use in the residential sector appears to be much larger in the long run than had been thought previously. We obtained an estimate of this elasticity of about −1.10. This is at the upper end of the range of estimates found by others, and certainly higher than the consensus range of estimates usually used for policy analysis and forecasting in the U.S.[6]

We also found considerable variation in price elasticities for individual fuels. We found own-price elasticities (total) for solid and liquid fuel to fall within the range of −1 and −1.25, those for natural gas to be about −1.7, and those for electricity to be between 0 and −0.4. Some of the own-price elasticity estimates for electricity are positive or unreasonably low (and the cross-price elasticity estimates are of the wrong sign), although they should be lower than those for other fuels. This may be due to income rather than price effects (the fuel model is constrained a priori to be homothetic), or from spurious estimates of one or more time trend parameters resulting from monotonic changes in particular prices and shares. Price elasticity estimates for the other fuels, however, are quite

6. A typical consensus estimate of the own-price elastictity of energy demand for purposes of policy analysis might be −0.25. See, for example, the review of U.S. energy policy by Hall and Pindyck (75).

reasonable, and we have also found them to be robust with respect to changes in the estimation time bounds or the particular pooling.

We have also seen that there is some evidence for pooling the U.S. and Canada separately when estimating the fuel-share model. If this is done, we find that the elasticities for liquid fuel and gas are still about −1 to −1.25 and about −1.7, respectively, for the European countries but only half as large for the U.S. and Canada. This is reasonable if one believes that price elasticities for necessities such as fuel become smaller as incomes rise (and there is certainly evidence that this is the case). On the other hand, another explanation for this difference in elasticities is that estimation with the U.S.-Canadian sample, for which more of the variation is timewise, may have elicited shorter-run elasticities. Pooling the U.S. and Canada separately yields elasticities for electricity near −1 for these countries but near 0 for some of the European countries, and these latter estimates are outside the range of most other studies. In addition, pooling the U.S. and Canada separately gives solid fuel elasticities that vary widely across countries and across time, but this is due largely to the fact that the share of solid fuel has become very small in most countries in recent years.

Let us now compare our results to those obtained by others. In table 4.12 we present a survey of recent estimates of residential energy demand elasticities. We can compare these estimates with those that we have obtained to determine whether there is any consensus on price elasticities, and how and why our estimates might differ from those of others.

Looking first at estimates of the long-run own-price elasticity of total energy use, we find a range extending from −0.28 to −1.70. However, only Nordhaus (123) obtained elasticities greater in magnitude than −1. An unweighted average of the seven studies other than that of Nordhaus gives an elasticity of −0.43, so that our estimate of about −1.1 would seem high. On the other hand, most of these other studies are based on time-series data for a single country (usually the U.S.), and thus are more likely to have captured short-run rather than long-run elasticities.

Most of the estimates of the long-run income elasticity of total energy use do not differ very much from unity, and this is supportive of our having found that the restriction of homotheticity can be accepted for the consumption expenditure model. Joskow and Baughman (102) obtain an elasticity of −0.6, and Nelson (122) obtains an elasticity of 0.27 (both for the U.S.), but these studies are based on time-series data for a single country and again are more likely to have captured short-run elasticities. There thus seems to be a consensus that the long-run income elasticity is about 1.

There is much less agreement on own-price elasticities of individual fuels. Elasticities for natural gas and fuel oil range from −0.33 to −1.89, although most are larger in magnitude than −1. Our elasticity estimates for oil (about −0.6 to −1.2) tend to be in the middle of this range, while our estimates for natural gas (about −0.9 to −1.8) are at the higher end. Elasticity estimates for electricity also vary considerably, ranging from −0.34 (Canada, see Fuss and Waverman, 62) to −1.2 (U.S., see Halvorsen, 76, and Mount, Chapman, and Tyrrell, 121). Notably, the one estimate for a European country (Rødseth and Strøm, 145) is at the low end of the spectrum. This is consistent with our finding that the estimated elasticity of electricity demand is larger in the U.S. and Canada when these countries are pooled separately from Europe. It is also reasonable since we expect that in the U.S. and Canada there is a greater discretionary use of electricity for such purposes as heating vacation homes or heating homes in warmer climates.

The main discrepancy between our results and those of others is that we have obtained elasticity estimates that are generally larger. This is true for aggregate energy use, and also (with the exception of electricity) for individual fuels. A large reason for this difference is probably our use of pooled international data, which, as we have argued, is more likely to elicit long-run elasticities. Another reason, however, may be our use of a general (that is, translog) functional form to model demand.

Although we expect that our elasticity estimates (or at least those that resulted from pooling all of the countries in our sample together) pertain to the long run, our results tell us nothing about how long the long run is, and in particular what the adjustment speeds of the various elasticities are. Based on depreciation rate data for household heating appliances, we might expect the median lag (or half-life) for the individual fuel elasticities to be around six to nine years. However, it is more difficult to say what the median lag should be for elasticities of aggregate energy demand because these will depend in part on consumption habits that are not directly tied to appliance purchases. The determination of these adjustment speeds requires the estimation of demand models that are explicitly dynamic. We have outlined some dynamic versions of the translog model in chapter 2, but we have not estimated them, and as we show in the appendix to this chapter, the estimation of dynamic logit models of fuel choice did not prove fruitful.

Our results indicate that long-run price elasticities of residential energy demand are larger than had been suggested by earlier work. This means that the use of taxes or other policies to raise retail fuel prices (such as deregulation in the U.S. and Canada) can be a very effective means of reducing or limiting energy consumption, given that enough time is allowed to pass to let demand adjust fully. In the past arguments against the use of such policies have often been based on low consensus estimates of demand elasticities. Hopefully energy policy in the future will be better guided by the recognition that long-run demand elasticities are in fact quite large.

## Appendix: Logit Models of Residential Fuel Shares

As an alternative to the translog model we have also estimated several versions of static and dynamic logit models to describe the dependence of fuel shares on prices, income, and temperature. The decision functions in these models are linear or logarithmic functions of relative fuel prices $\tilde{P}_i$ (the price of fuel $i$ divided by the price of energy), per capita income $Y$, temperature $T$, and, in the dynamic models, lagged shares. Recall from equation (2.36) that this leads to a set of three equations that must be estimated simultaneously, since certain coefficients are constrained to be the same across equations. We therefore use iterative Zellner estimation to estimate all of the models.

Our static logit models are of the general form

$$\log (S_i/S_4) = \sum_{k=1}^{9} a_{i4k}D_k + b_i\tilde{P}_i - b_4\tilde{P}_4 + c_{i4}Y + d_{i4}T, \qquad i = 1, 2, 3, \tag{4.4}$$

where $a_{i4k} = a_{ik} - a_{4k}$, $c_{i4} = c_i - c_4$, and $d_{i4} = d_i - d_4$, as in equation (2.34). The $D_k$ are country dummy variables (countries are ordered alphabetically), and the fuels are ordered (1) liquid, (2) solid, (3) gas, and (4) electricity. Parameter estimates for four alternative versions of the model are shown in table 4.13 ($t$-statistics are in parentheses). These versions differ in terms of which explanatory variables are included in the decision function, and whether the decision function is linear or log-linear.

Unfortunately, the estimation results are not encouraging. Own-price elasticities are determined by the coefficients $b_i$, and for three of the models two or more of these are positive. The version of the model that gives the most sensible results is that shown in column 3, in which the decision function is linear in relative prices and income, price dummy variables are included for the shares of solid fuel in Canada and the U.S. and for the shares of natural gas in Norway and West Germany,[7] and the temperature variable is not included. Even here, however, $b_4$ (the coefficient determining the own-price elasticity for electricity) is positive (but insignificant), and only $b_3$ is significant at the 5 percent level.

We have calculated price and income elasticities for this version of the static logit model, but we do not present them in detail here. Only the own-price elasticities for natural gas are reasonable; these are in the range −0.50 to −1.90. The other own-price elasticities are insignificantly different from 0. Income elasticities are large and negative for solid fuel (between −1.5 and −5.5). These do not seem reasonable, although small negative income elasticities might be if consumers shift to cleaner and more convenient fuels as their incomes rise. Income elasticities for other fuels range from 0 to 2.5, but in many cases are statistically insignificant.

The dynamic version of the logit model is based on the assumption that the choice of fuels this period depends on the relative

7. The price dummy variables for solid fuel in Canada and the U.S. are $CNSD$ and $USSD$, and for gas in Norway and West Germany are $NRGD$ and $WGGD$. There is virtually no solid fuel used in the residential sectors of Canada and the U.S., and little or no gas used in the residential sectors of Norway and West Germany. This is not because prices are too high, but because in Canada and the U.S. other fuels are readily available that are cleaner and more convenient, and in Norway and West Germany the extremely limited supplies of gas are not made available to residential consumers. Note that the price dummy variables are indeed highly significant.

shares last period, as well as this period's prices and income. The dependence on past shares is intended to incorporate both habit formation and stock adjustment effects. It leads to equations of the form

$$\log (S_i/S_4) = \sum_{k=1}^{9} a_{i4k}D_k + b_i\tilde{P}_i - b_4\tilde{P}_4 + c_{i4}Y$$

$$+ \lambda_i S_{i,t-1} - \lambda_4 S_{4,t-1}, \qquad i = 1, 2, 3. \qquad (4.5)$$

Again this is not a Koyck adjustment model; the coefficients $\lambda_i$ can be greater than 1 (although we would expect them to be positive), and in general a change in price will not lead to geometrically declining changes in shares over time.

Three versions of this model were estimated. The first is identical to the static model of column 3 in table 4.13, except that it includes lagged share terms. The second is identical to the static model of column 4 but with lagged shares, and the third is the same as the first but uses actual rather than relative prices.

The estimation results are again discouraging. None of the $b_i$ coefficients are significant at the 5 percent level, and at least one is positive in each model. This inability of both the static and dynamic versions of this model to pick up price effects may be due to its restrictive form. All cross-price elasticities for a given own price are constrained to be equal, and parameter estimates become inefficient when shares become very small.

**Table 4.1**
Parameter estimates of the translog energy price aggregator[a]

| Parameters | (1) Four fuels, nine countries | (2) Four fuels, nine countries: regional dummy variables | (3) Four fuels, seven countries: regional dummy variables | (4) Three fuels, U.S. and Canada: regional dummy variables |
|---|---|---|---|---|
| $\gamma_1$ | −0.0438 (0.0380) | | | |
| $\gamma_1 D_1$ | | 0.0657 (0.0488) | 0.0450 (0.0536) | |
| $\gamma_1 D_2$ | | −0.0454 (0.0360) | | |
| $\gamma_1 D_3$ | | −0.0051 (0.0457) | −0.0193 (0.0501) | |
| $\gamma_1 D_4$ | | −0.2445 (0.0493) | −0.2634 (0.0545) | |
| $\gamma_1 D_5$ | | −0.0191 (0.0405) | −0.0327 (0.0430) | |
| $\gamma_1 D_6$ | | −0.0872 (0.0308) | −0.0873 (0.0349) | |
| $\gamma_1 D_7$ | | −0.0104 (0.0450) | −0.0338 (0.0517) | |
| $\gamma_1 D_8$ | | −0.0186 (0.0452) | | |
| $\gamma_1 D_9$ | | −0.2516 (0.0478) | −0.2653 (0.0532) | |
| $\gamma_2$ | 0.2517 (0.0229) | | | |
| $\gamma_2 D_1$ | | 0.2922 (0.0399) | 0.3359 (0.0426) | |
| $\gamma_2 D_2$ | | 0.5236 (0.0282) | | 0.2902 (0.0226) |
| $\gamma_2 D_3$ | | 0.2673 (0.0404) | 0.2991 (0.0434) | |
| $\gamma_2 D_4$ | | 0.2498 (0.0408) | 0.2867 (0.0439) | |
| $\gamma_2 D_5$ | | 0.3546 (0.0352) | 0.3921 (0.0372) | |
| $\gamma_2 D_6$ | | 0.1469 (0.0260) | 0.1339 (0.0290) | |
| $\gamma_2 D_7$ | | 0.1547 (0.0302) | 0.1906 (0.0336) | |
| $\gamma_2 D_8$ | | 0.5336 (0.0356) | | 0.1949 (0.0279) |
| $\gamma_2 D_9$ | | 0.2923 (0.0416) | 0.3166 (0.0453) | |
| $\gamma_3$ | 0.0824 (0.0299) | | | |
| $\gamma_3 D_1$ | | 0.5983 (0.0411) | 0.7015 (0.0433) | |
| $\gamma_3 D_2$ | | 0.2865 (0.0363) | | 0.0665 (0.0234) |
| $\gamma_3 D_3$ | | 0.6416 (0.0393) | 0.7421 (0.0420) | |
| $\gamma_3 D_4$ | | 0.7065 (0.0414) | 0.8091 (0.0442) | |
| $\gamma_3 D_5$ | | 0.6367 (0.0346) | 0.7269 (0.0361) | |
| $\gamma_3 D_6$ | | 0.1947 (0.0275) | 0.2251 (0.0294) | |
| $\gamma_3 D_7$ | | 0.5380 (0.0327) | 0.5957 (0.0345) | |
| $\gamma_3 D_8$ | | 0.3492 (0.0494) | | 0.1335 (0.0327) |
| $\gamma_3 D_9$ | | 0.5767 (0.0410) | 0.6753 (0.0444) | |

[a] Country dummy variables are numbered alphabetically.

| Parameters | (1) Four fuels, nine countries | (2) Four fuels, nine countries: regional dummy variables | (3) Four fuels, seven countries: regional dummy variables | (4) Three fuels, U.S. and Canada: regional dummy variables |
|---|---|---|---|---|
| $\gamma_4$ | 0.7097 (0.0403) | | | |
| $\gamma_4 D_1$ | | 0.0439 (0.0634) | −0.0820 (0.0429) | |
| $\gamma_4 D_2$ | | 0.2353 (0.0548) | | 0.6433 (0.0291) |
| $\gamma_4 D_3$ | | 0.0963 (0.0544) | −0.0220 (0.0389) | |
| $\gamma_4 D_4$ | | 0.2881 (0.0426) | 0.1680 (0.0419) | |
| $\gamma_4 D_5$ | | 0.0279 (0.0365) | −0.0860 (0.0356) | |
| $\gamma_4 D_6$ | | 0.7456 (0.0199) | 0.7282 (0.0203) | |
| $\gamma_4 D_7$ | | 0.3177 (0.0348) | 0.2734 (0.0347) | |
| $\gamma_4 D_8$ | | 0.1352 (0.0485) | | 0.6715 (0.0377) |
| $\gamma_4 D_9$ | | 0.3829 (0.0365) | −0.7266 (0.0389) | |
| $\gamma_0$ | 9.066 | 9.115 | 9.773 | 9.161 |
| $\gamma_{11}$ | −0.3701 (0.0424) | −0.2596 (0.0450) | −0.2797 (0.0529) | |
| $\gamma_{12}$ | 0.1257 (0.0239) | 0.0007 (0.0292) | 0.0105 (0.0342) | |
| $\gamma_{13}$ | 0.0915 (0.0169) | 0.1537 (0.0271) | 0.1469 (0.0292) | |
| $\gamma_{14}$ | 0.1529 (0.0265) | 0.1051 (0.0244) | 0.1222 (0.0264) | |
| $\gamma_{21}$ | 0.1257 (0.0239) | 0.0007 (0.0292) | 0.0105 (0.0342) | |
| $\gamma_{22}$ | −0.0292 (0.0179) | −0.0032 (0.0278) | −0.0118 (0.0320) | 0.0555 (0.0180) |
| $\gamma_{23}$ | −0.0933 (0.0108) | 0.1032 (0.0212) | 0.1459 (0.0397) | −0.1062 (0.0117) |
| $\gamma_{24}$ | −0.0032 (0.0150) | −0.1007 (0.0178) | −0.1445 (0.0192) | 0.0507 (0.0163) |
| $\gamma_{31}$ | 0.0915 (0.0169) | 0.1537 (0.0271) | 0.1469 (0.0292) | |
| $\gamma_{32}$ | −0.0933 (0.0108) | 0.1031 (0.0212) | 0.1459 (0.0237) | −0.1062 (0.0117) |
| $\gamma_{33}$ | −0.0747 (0.0152) | −0.1241 (0.0316) | −0.1181 (0.0325) | 0.0303 (0.0194) |
| $\gamma_{34}$ | 0.0764 (0.0160) | −0.1328 (0.0229) | −0.1747 (0.0227) | 0.0759 (0.0157) |
| $\gamma_{41}$ | 0.1529 (0.0265) | 0.1051 (0.0244) | 0.1222 (0.0264) | |
| $\gamma_{42}$ | −0.0032 (0.0150) | −0.1007 (0.0178) | −0.1445 (0.0192) | 0.0507 (0.0163) |
| $\gamma_{43}$ | 0.0764 (0.0160) | −0.1328 (0.0229) | −0.1747 (0.0227) | 0.0759 (0.0157) |
| $\gamma_{44}$ | −0.2260 (0.0283) | 0.1283 (0.0262) | 0.1970 (0.0254) | −0.1266 (0.0203) |
| **RSQ** | | | | |
| Equation 1 | 0.374 | 0.795 | 0.729 | |
| Equation 2 | 0.435 | 0.711 | 0.518 | 0.674 |
| Equation 3 | 0.142 | 0.746 | 0.782 | 0.944 |

**Table 4.2**
Energy price indices (U.S. = 1 in 1970)

| | (1) Four fuels for all countries | | | |
|---|---|---|---|---|
| | Belgium | | Canada | |
| | (1) | (2) | (1) | (2) |
| 1955 | — | — | — | — |
| 1956 | — | — | — | — |
| 1957 | — | — | — | — |
| 1958 | 2.2344 | 2.2902 | 0.9979 | 0.9730 |
| 1959 | 2.1719 | 2.1979 | 1.0060 | 0.9863 |
| 1960 | 2.2227 | 2.2473 | 0.9867 | 0.9601 |
| 1961 | 2.1314 | 2.1422 | 0.9750 | 0.9466 |
| 1962 | 2.1230 | 2.1377 | 0.9580 | 0.9341 |
| 1963 | 2.0643 | 2.0681 | 0.9410 | 0.9275 |
| 1964 | 2.1289 | 2.1411 | 0.9243 | 0.9057 |
| 1965 | 2.1164 | 2.1247 | 0.9044 | 0.8858 |
| 1966 | 1.9856 | 2.0003 | 0.8615 | 0.8492 |
| 1967 | 1.9263 | 1.9349 | 0.8754 | 0.8409 |
| 1968 | 1.8922 | 1.9073 | 0.8905 | 0.8499 |
| 1969 | 1.8523 | 1.8649 | 0.8763 | 0.8371 |
| 1970 | 1.9315 | 1.9315 | 0.8685 | 0.8244 |
| 1971 | 2.0218 | 2.0170 | 0.8771 | 0.8432 |
| 1972 | 1.8737 | 1.8520 | 0.8544 | 0.8325 |
| 1973 | 1.8229 | 1.7725 | 0.8493 | 0.8483 |
| 1974 | 1.7878 | 1.7490 | 1.0105 | 1.0324 |

| (2) Bridging method for three fuels of U.S. and Canada | | | | | |
|---|---|---|---|---|---|
| France | | Italy | | Netherlands | |
| (1) | (2) | (1) | (2) | (1) | (2) |
| — | — | — | — | 1.9248 | 1.9121 |
| — | — | — | — | 1.9585 | 1.9391 |
| — | — | — | — | 1.9853 | 1.9725 |
| 2.6068 | 2.6279 | — | — | 1.9747 | 1.9810 |
| 2.6541 | 2.6594 | — | — | 1.9404 | 1.9519 |
| 2.6727 | 2.6864 | 3.4913 | 3.5180 | 1.8876 | 1.9000 |
| 2.6431 | 2.6561 | 3.3705 | 3.4002 | 1.7850 | 1.7970 |
| 2.5045 | 2.5042 | 3.2250 | 3.2484 | 1.7264 | 1.7228 |
| 2.4536 | 2.4566 | 3.0037 | 3.0198 | 1.7030 | 1.6922 |
| 2.4551 | 2.4657 | 2.9525 | 2.9746 | 1.6783 | 1.6723 |
| 2.3630 | 2.3741 | 2.9405 | 2.9636 | 1.5520 | 1.5507 |
| 2.3431 | 2.3700 | 2.8788 | 2.9075 | 1.5588 | 1.5855 |
| 2.3381 | 2.3800 | 2.7040 | 2.6913 | 1.5365 | 1.5535 |
| 2.2264 | 2.2341 | 2.6092 | 2.5928 | 1.5238 | 1.5334 |
| 2.1808 | 2.1852 | 2.4359 | 2.4048 | 1.4067 | 1.4080 |
| 2.1790 | 2.2074 | 2.3128 | 2.2801 | 1.4082 | 1.4047 |
| 2.0735 | 2.0861 | 2.5631 | 2.5457 | 1.3974 | 1.3920 |
| 2.1048 | 2.0719 | 2.4661 | 2.4439 | 1.3054 | 1.3013 |
| 1.9570 | 1.9174 | 2.4280 | 2.3978 | 1.2894 | 1.2578 |
| 1.8190 | 1.7645 | 2.2953 | 2.2734 | 1.3352 | 1.2812 |

**Table 4.2**
(continued)

| | (1) Four fuels for all countries | | | |
|---|---|---|---|---|
| | Norway | | U.K. | |
| | (1) | (2) | (1) | (2) |
| 1955 | — | — | — | — |
| 1956 | — | — | — | — |
| 1957 | — | — | — | — |
| 1958 | — | — | — | — |
| 1959 | — | — | — | — |
| 1960 | — | — | 1.5776 | 1.5579 |
| 1961 | — | — | 1.5561 | 1.5320 |
| 1962 | — | — | 1.6193 | 1.5972 |
| 1963 | — | — | 1.6501 | 1.6299 |
| 1964 | 1.1937 | 1.2013 | 1.6712 | 1.6584 |
| 1965 | 1.1455 | 1.1505 | 1.7199 | 1.7190 |
| 1966 | 1.1287 | 1.1349 | 1.7157 | 1.7203 |
| 1967 | 1.1579 | 1.1672 | 1.7820 | 1.7897 |
| 1968 | 1.1515 | 1.1597 | 1.8508 | 1.8680 |
| 1969 | 1.1308 | 1.1300 | 1.7315 | 1.7466 |
| 1970 | 0.9742 | 0.9548 | 1.6798 | 1.6958 |
| 1971 | 1.1702 | 1.1611 | 1.7092 | 1.7320 |
| 1972 | 1.1134 | 1.1006 | 1.7616 | 1.7780 |
| 1973 | 1.1105 | 1.1033 | 1.7050 | 1.6973 |
| 1974 | 1.2660 | 1.2772 | 1.6272 | 1.6281 |

| (2) Bridging method for three fuels of U.S. and Canada | | | |
|---|---|---|---|
| U.S. | | West Germany | |
| (1) | (2) | (1) | (2) |
| — | — | — | — |
| — | — | 3.7413 | 3.8100 |
| — | — | 3.7579 | 3.7949 |
| — | — | 3.7123 | 3.8358 |
| 1.3976 | 1.0934 | 3.6608 | 3.7903 |
| 1.3657 | 1.0960 | 3.3493 | 3.4658 |
| 1.3430 | 1.1060 | 3.2496 | 3.3752 |
| 1.3149 | 1.0879 | 3.2551 | 3.3444 |
| 1.2849 | 1.0847 | 3.1409 | 3.2135 |
| 1.2604 | 1.1141 | 2.9953 | 3.0864 |
| 1.2173 | 1.0996 | 2.8751 | 2.9644 |
| 1.1764 | 1.0912 | 2.7529 | 2.8463 |
| 1.1387 | 1.0728 | 2.6015 | 2.6198 |
| 1.0869 | 1.0495 | 2.7237 | 2.7226 |
| 1.0365 | 1.0183 | 2.5262 | 2.5289 |
| 1.0000 | 1.0000 | 2.4103 | 2.3912 |
| 1.0097 | 1.0215 | 2.3652 | 2.3108 |
| 1.0427 | 1.0901 | 2.2262 | 2.1516 |
| 1.1011 | 1.3109 | 2.5113 | 2.4245 |
| 1.1557 | 1.3329 | 2.6600 | 2.5699 |

**Table 4.3**
Parameter estimates of consumption

| Parameter[a] | (1) Nonhomothetic, nonstationary | (2) Homothetic, nonstationary, 1960–1974 (preferred model) | (3) Homothetic, stationary, 1960–1974 |
|---|---|---|---|
| $\alpha_1 D_1$ | −0.0697 (0.1363) | −0.0828 (0.0081) | −0.0755 (0.0038) |
| $\alpha_1 D_2$ | −0.0789 (0.1389) | −0.0894 (0.0049) | −0.0845 (0.0028) |
| $\alpha_1 D_3$ | −0.0883 (0.1334) | −0.0932 (0.0082) | −0.0848 (0.0055) |
| $\alpha_1 D_4$ | −0.0867 (0.1263) | −0.0917 (0.0095) | −0.0849 (0.0064) |
| $\alpha_1 D_5$ | −0.1123 (0.1310) | −0.1080 (0.0067) | −0.1010 (0.0041) |
| $\alpha_1 D_6$ | −0.1396 (0.1418) | −0.1247 (0.0063) | −0.1231 (0.0047) |
| $\alpha_1 D_7$ | −0.0756 (0.1393) | −0.0855 (0.0079) | −0.0797 (0.0050) |
| $\alpha_1 D_8$ | −0.0793 (0.1422) | −0.0908 (0.0053) | −0.0840 (0.0016) |
| $\alpha_1 D_9$ | −0.1184 (0.1347) | −0.1107 (0.0103) | −0.1034 (0.0068) |
| $\alpha_2 D_1$ | −0.1388 (0.2338) | −0.1388 (0.0096) | −0.1453 (0.0050) |
| $\alpha_2 D_2$ | −0.0304 (0.2612) | −0.0764 (0.0064) | −0.0700 (0.0041) |
| $\alpha_2 D_3$ | −0.0352 (0.2529) | −0.0764 (0.0100) | −0.0811 (0.0075) |
| $\alpha_2 D_4$ | −0.0297 (0.2412) | −0.0708 (0.0111) | −0.0776 (0.0083) |
| $\alpha_2 D_5$ | −0.1153 (0.2317) | −0.1255 (0.0084) | −0.1238 (0.0060) |
| $\alpha_2 D_6$ | −0.0610 (0.2421) | −0.0941 (0.0082) | −0.0816 (0.0072) |
| $\alpha_2 D_7$ | −0.0232 (0.2634) | −0.0696 (0.0095) | −0.0681 (0.0070) |
| $\alpha_2 D_8$ | −0.1617 (0.2610) | −0.1540 (0.0072) | −0.1567 (0.0028) |
| $\alpha_2 D_9$ | −0.1250 (0.2354) | −0.1300 (0.0119) | −0.1355 (0.0089) |
| $\alpha_3 D_1$ | −0.3906 (0.3128) | −0.3825 (0.0104) | −0.2866 (0.0073) |
| $\alpha_3 D_2$ | −0.2926 (0.3356) | −0.3226 (0.0086) | −0.2395 (0.0063) |
| $\alpha_3 D_3$ | −0.4077 (0.3215) | −0.3906 (0.0112) | −0.2934 (0.0085) |
| $\alpha_3 D_4$ | −0.5118 (0.3627) | −0.4554 (0.0119) | −0.3741 (0.0102) |
| $\alpha_3 D_5$ | −0.3763 (0.3105) | −0.3707 (0.0102) | −0.2728 (0.0072) |
| $\alpha_3 D_6$ | −0.4452 (0.3365) | −0.4125 (0.0100) | −0.3462 (0.0095) |
| $\alpha_3 D_7$ | −0.4355 (0.3378) | −0.4073 (0.0107) | −0.3178 (0.0082) |
| $\alpha_3 D_8$ | −0.3091 (0.3423) | −0.3358 (0.0088) | −0.2416 (0.0056) |
| $\alpha_3 D_9$ | −0.3454 (0.3101) | −0.3540 (0.0126) | −0.2622 (0.0103) |
| $\alpha_4 D_1$ | −0.0874 (0.1528) | −0.0791 (0.0084) | −0.1179 (0.0045) |
| $\alpha_4 D_2$ | −0.1661 (0.1775) | −0.1224 (0.0049) | −0.1512 (0.0032) |
| $\alpha_4 D_3$ | −0.1029 (0.1537) | −0.0886 (0.0083) | −0.1187 (0.0070) |
| $\alpha_4 D_4$ | −0.0974 (0.1451) | −0.0871 (0.0096) | −0.1203 (0.0077) |
| $\alpha_4 D_5$ | −0.0468 (0.1716) | −0.0534 (0.0066) | −0.0847 (0.0050) |
| $\alpha_4 D_6$ | −0.1135 (0.1535) | −0.0920 (0.0065) | −0.1240 (0.0055) |
| $\alpha_4 D_7$ | −0.1347 (0.1605) | −0.1069 (0.0078) | −0.1415 (0.0058) |

[a] Consumption categories are 1 = apparel, 2 = durables, 3 = food, 4 = transportation and communication, 5 = energy, 6 = all other. The country dummy variables are numbered alphabetically: 1 = Belgium, 2 = Canada, 3 = France, 4 = Italy, 5 = Netherlands, 6 = Norway, 7 = U.K., 8 = U.S., 9 = West Germany.

| (4) Homothetic, nonstationary, energy separability, 1960–1974 | (5) Homothetic, nonstationary, with additivity, 1960–1974 | (6) Homothetic, nonstationary, data at four-year intervals, 1962–1973 |
|---|---|---|
| −0.0875 (0.0051) | −0.1079 (0.0027) | −0.1019 (0.0084) |
| −0.0911 (0.0045) | −0.1110 (0.0028) | −0.0986 (0.0049) |
| −0.0989 (0.0050) | −0.1231 (0.0027) | −0.1160 (0.0090) |
| −0.0979 (0.0055) | −0.1187 (0.0027) | −0.1206 (0.0107) |
| −0.1121 (0.0051) | −0.1376 (0.0027) | −0.1239 (0.0070) |
| −0.1268 (0.0056) | −0.1422 (0.0031) | −0.1439 (0.0072) |
| −0.0901 (0.0056) | −0.1143 (0.0027) | −0.1064 (0.0086) |
| −0.0934 (0.0041) | −0.1082 (0.0027) | −0.0988 (0.0051) |
| −0.1173 (0.0063) | −0.1438 (0.0028) | −0.1402 (0.0114) |
| −0.1422 (0.0064) | −0.1371 (0.0047) | −0.1338 (0.0095) |
| −0.0776 (0.0060) | −0.0845 (0.0049) | −0.0768 (0.0062) |
| −0.0811 (0.0064) | −0.0926 (0.0047) | −0.0743 (0.0105) |
| −0.0761 (0.0061) | −0.0748 (0.0047) | −0.0705 (0.0119) |
| −0.1287 (0.0068) | −0.1426 (0.0047) | −0.1249 (0.0085) |
| −0.0969 (0.0075) | −0.1033 (0.0053) | −0.0976 (0.0088) |
| −0.0737 (0.0070) | −0.0826 (0.0047) | −0.0692 (0.0098) |
| −0.1553 (0.0062) | −0.1529 (0.0047) | −0.1531 (0.0069) |
| −0.1357 (0.0071) | −0.1392 (0.0048) | −0.1285 (0.0126) |
| −0.4013 (0.0082) | −0.4185 (0.0072) | −0.3874 (0.0123) |
| −0.3259 (0.0087) | −0.3516 (0.0074) | −0.3167 (0.0092) |
| −0.4133 (0.0088) | −0.4423 (0.0072) | −0.3905 (0.0137) |
| −0.4835 (0.0081) | −0.5074 (0.0072) | −0.4582 (0.0156) |
| −0.3837 (0.0095) | −0.4172 (0.0072) | −0.3661 (0.0117) |
| −0.4201 (0.0099) | −0.4548 (0.0081) | −0.4066 (0.0127) |
| −0.4246 (0.0093) | −0.4596 (0.0072) | −0.4034 (0.0131) |
| −0.3433 (0.0083) | −0.3522 (0.0072) | −0.3432 (0.0094) |
| −0.3820 (0.0091) | −0.4151 (0.0073) | −0.3501 (0.0162) |
| −0.0660 (0.0054) | −0.0623 (0.0030) | −0.0717 (0.0083) |
| −0.1184 (0.0045) | −0.1085 (0.0031) | −0.1198 (0.0050) |
| −0.0743 (0.0048) | −0.0600 (0.0030) | −0.0789 (0.0091) |
| −0.0703 (0.0049) | −0.0591 (0.0030) | −0.0734 (0.0106) |
| −0.0438 (0.0049) | −0.0289 (0.0030) | −0.0469 (0.0070) |
| −0.0858 (0.0058) | −0.0712 (0.0034) | −0.0843 (0.0074) |
| −0.0949 (0.0053) | −0.0797 (0.0030) | −0.0973 (0.0084) |

**Table 4.3**
(continued)

| Parameter | (1) Nonhomothetic, nonstationary | (2) Homothetic, nonstationary, 1960-1974 (preferred model) | (3) Homothetic, stationary, 1960-1974 |
|---|---|---|---|
| $\alpha_4 D_8$ | −0.0400 (0.2618) | −0.0001 (0.0060) | −0.0329 (0.0022) |
| $\alpha_4 D_9$ | −0.1263 (0.1554) | −0.1039 (0.0102) | −0.1408 (0.0080) |
| $\alpha_5 D_1$ | −0.0559 (0.1196) | −0.0500 (0.0069) | −0.0561 (0.0037) |
| $\alpha_5 D_2$ | −0.0163 (0.1223) | −0.0271 (0.0036) | −0.0273 (0.0020) |
| $\alpha_5 D_3$ | −0.0293 (0.1180) | −0.0352 (0.0073) | −0.0411 (0.0051) |
| $\alpha_5 D_4$ | −0.0186 (0.1128) | −0.0283 (0.0085) | −0.0339 (0.0061) |
| $\alpha_5 D_5$ | −0.0554 (0.1194) | −0.0509 (0.0054) | −0.0539 (0.0033) |
| $\alpha_5 D_6$ | −0.0148 (0.1194) | −0.0265 (0.0048) | −0.0225 (0.0036) |
| $\alpha_5 D_7$ | −0.0474 (0.1214) | −0.0460 (0.0065) | −0.0482 (0.0043) |
| $\alpha_5 D_8$ | −0.0282 (0.1242) | −0.0331 (0.0044) | −0.0379 (0.0015) |
| $\alpha_5 D_9$ | −0.0561 (0.1220) | −0.0515 (0.0089) | −0.0569 (0.0063) |
| $\alpha_6 D_1$ | −0.2576 (0.2114) | −0.2666 (0.0128) | −0.3182 (0.0087) |
| $\alpha_6 D_2$ | −0.4157 (0.3151) | −0.3618 (0.0125) | −0.4272 (0.0075) |
| $\alpha_6 D_3$ | −0.3366 (0.2871) | −0.3156 (0.0141) | −0.3807 (0.0097) |
| $\alpha_6 D_4$ | −0.2558 (0.2724) | −0.2664 (0.0143) | −0.3089 (0.0117) |
| $\alpha_6 D_5$ | −0.2939 (0.2014) | −0.2912 (0.0140) | −0.3636 (0.0083) |
| $\alpha_6 D_6$ | −0.2259 (0.1917) | −0.2499 (0.0139) | −0.3022 (0.0102) |
| $\alpha_6 D_7$ | −0.2836 (0.2201) | −0.2844 (0.0141) | −0.3443 (0.0091) |
| $\alpha_6 D_8$ | −0.3817 (0.2122) | −0.3860 (0.0114) | −0.4467 (0.0069) |
| $\alpha_6 D_9$ | −0.2288 (0.1808) | −0.2495 (0.0153) | −0.3009 (0.0114) |
| $\beta_{11}$ | 0.0777 (0.1090) | 0.0501 (0.0113) | −0.0542 (0.0099) |
| $\beta_{12}$ | −0.0226 (0.0353) | −0.0123 (0.0093) | −0.0185 (0.0074) |
| $\beta_{13}$ | 0.0650 (0.0952) | 0.0386 (0.0084) | −0.0570 (0.0078) |
| $\beta_{14}$ | −0.0330 (0.0427) | −0.0217 (0.0085) | −0.0312 (0.0082) |
| $\beta_{15}$ | −0.0043 (0.0108) | −0.0044 (0.0056) | −0.0075 (0.0046) |
| $\beta_{16}$ | −0.0744 (0.0963) | −0.0502 (0.0091) | −0.0539 (0.0082) |
| $\beta_{21}$ | −0.0226 (0.0353) | −0.0123 (0.0093) | −0.0185 (0.0074) |
| $\beta_{22}$ | 0.0958 (0.1379) | 0.0538 (0.0147) | 0.0796 (0.0137) |
| $\beta_{23}$ | 0.0654 (0.0975) | 0.0426 (0.0119) | 0.0192 (0.0110) |
| $\beta_{24}$ | −0.0464 (0.0608) | −0.0272 (0.0105) | −0.0151 (0.0096) |
| $\beta_{25}$ | −0.0017 (0.0137) | −0.0041 (0.0068) | 0.0084 (0.0059) |
| $\beta_{26}$ | −0.0786 (0.1031) | −0.0527 (0.0140) | −0.0737 (0.0131) |
| $\beta_{31}$ | 0.0650 (0.0952) | 0.0386 (0.0084) | 0.0570 (0.0078) |
| $\beta_{32}$ | 0.0654 (0.0975) | 0.0426 (0.0119) | 0.0192 (0.0110) |
| $\beta_{33}$ | 0.0272 (0.0741) | 0.0142 (0.0184) | 0.1718 (0.0210) |
| $\beta_{34}$ | −0.0058 (0.0186) | −0.0081 (0.0098) | −0.0242 (0.0108) |
| $\beta_{35}$ | −0.0381 (0.0482) | −0.0224 (0.0071) | −0.0416 (0.0062) |
| $\beta_{36}$ | −0.0858 (0.1007) | −0.0649 (0.0207) | −0.1823 (0.0221) |
| $\beta_{41}$ | −0.0030 (0.0427) | −0.0217 (0.0085) | −0.0312 (0.0082) |

| (4) Homothetic, nonstationary, energy separability, 1960–1974 | (5) Homothetic, nonstationary, with additivity, 1960–1974 | (6) Homothetic, nonstationary, data at four-year intervals, 1962–1973 |
|---|---|---|
| −0.0066 (0.0050) | 0.0057 (0.0030) | 0.0023 (0.0057) |
| −0.0865 (0.0056) | −0.0712 (0.0031) | −0.0909 (0.0111) |
| −0.0446 (0.0022) | −0.0446 (0.0022) | −0.0482 (0.0073) |
| −0.0227 (0.0023) | −0.0227 (0.0023) | −0.0240 (0.0035) |
| −0.0255 (0.0022) | −0.0255 (0.0022) | −0.0327 (0.0080) |
| −0.0216 (0.0022) | −0.0216 (0.0022) | −0.0272 (0.0096) |
| −0.0424 (0.0022) | −0.0424 (0.0022) | −0.0475 (0.0057) |
| −0.0243 (0.0025) | −0.0243 (0.0025) | −0.0252 (0.0054) |
| −0.0389 (0.0022) | −0.0389 (0.0022) | −0.0435 (0.0071) |
| −0.0290 (0.0022) | −0.0290 (0.0022) | −0.0817 (0.0048) |
| −0.0433 (0.0022) | −0.0433 (0.0022) | −0.0463 (0.0099) |
| −0.2580 (0.0117) | −0.2292 (0.0106) | −0.2567 (0.0127) |
| −0.3638 (0.0126) | −0.3214 (0.0109) | −0.3638 (0.0119) |
| −0.3064 (0.0128) | −0.2562 (0.0106) | −0.3074 (0.0145) |
| −0.2502 (0.0114) | −0.2181 (0.0106) | −0.2498 (0.0152) |
| −0.2891 (0.0139) | −0.2311 (0.0106) | −0.2904 (0.0140) |
| −0.2458 (0.0138) | −0.2040 (0.0210) | −0.2421 (0.0137) |
| −0.2774 (0.0134) | −0.2246 (0.0106) | −0.2800 (0.0141) |
| −0.3855 (0.0114) | −0.3633 (0.0106) | −0.3752 (0.0110) |
| −0.2349 (0.0129) | −0.1871 (0.0108) | −0.0243 (0.0160) |
| 0.0481 (0.0110) | 0 | 0.0330 (0.0111) |
| −0.0123 (0.0090) | 0 | −0.0154 (0.0094) |
| 0.0336 (0.0078) | 0 | −0.0610 (0.0115) |
| −0.0195 (0.0084) | 0 | −0.0189 (0.0083) |
| 0 | 0 | 0.0090 (0.0060) |
| −0.0498 (0.0090) | 0 | −0.0686 (0.0094) |
| −0.0123 (0.0090) | 0 | −0.0154 (0.0094) |
| 0.0502 (0.0146) | 0 | 0.0523 (0.0154) |
| 0.0423 (0.0117) | 0 | 0.0611 (0.0161) |
| −0.0271 (0.0104) | 0 | −0.0220 (0.0109) |
| 0 | 0 | −0.0082 (0.0070) |
| −0.0531 (0.0138) | 0 | −0.0677 (0.0151) |
| 0.0336 (0.0078) | 0 | −0.0610 (0.0115) |
| 0.0423 (0.0117) | 0 | 0.0611 (0.0161) |
| 0.0105 (0.0183) | 0 | 0.0013 (0.0277) |
| −0.0009 (0.0094) | 0 | −0.0217 (0.0133) |
| 0 | 0 | −0.0149 (0.0088) |
| −0.0855 (0.0202) | 0 | −0.0867 (0.0256) |
| −0.0195 (0.0084) | 0 | −0.0189 (0.0083) |

**Table 4.3**
(continued)

| Parameter[a] | (1) Nonhomothetic, nonstationary | (2) Homothetic, nonstationary, 1960-1974 (preferred model) | (3) Homothetic, stationary, 1960-1974 |
|---|---|---|---|
| $\beta_{42}$ | −0.0464 (0.0608) | −0.0272 (0.0105) | −0.0151 (0.0096) |
| $\beta_{43}$ | −0.0058 (0.0186) | −0.0081 (0.0098) | −0.0219 (0.0108) |
| $\beta_{44}$ | 0.0242 (0.0386) | 0.0161 (0.0134) | −0.0203 (0.0131) |
| $\beta_{45}$ | 0.0148 (0.0219) | 0.0128 (0.0064) | 0.0219 (0.0058) |
| $\beta_{46}$ | 0.0493 (0.0722) | 0.0281 (0.0103) | 0.0690 (0.0105) |
| $\beta_{51}$ | −0.0043 (0.0108) | −0.0044 (0.0056) | −0.0075 (0.0046) |
| $\beta_{52}$ | −0.0017 (0.0137) | −0.0041 (0.0068) | 0.0084 (0.0059) |
| $\beta_{53}$ | −0.0381 (0.0482) | −0.0224 (0.0071) | −0.0416 (0.0062) |
| $\beta_{54}$ | 0.0148 (0.0219) | 0.0128 (0.0064) | 0.0219 (0.0058) |
| $\beta_{55}$ | 0.0075 (0.0138) | 0.0041 (0.0060) | 0.0102 (0.0051) |
| $\beta_{56}$ | 0.0250 (0.0387) | 0.0139 (0.0076) | 0.0085 (0.0074) |
| $\beta_{61}$ | −0.0744 (0.0963) | −0.0502 (0.0091) | −0.0539 (0.0082) |
| $\beta_{62}$ | −0.0786 (0.1031) | −0.0527 (0.0140) | −0.0737 (0.0131) |
| $\beta_{63}$ | −0.0858 (0.1007) | −0.0649 (0.0207) | −0.1823 (0.0221) |
| $\beta_{64}$ | 0.0493 (0.0722) | 0.0281 (0.0103) | 0.0690 (0.0105) |
| $\beta_{65}$ | 0.0250 (0.0387) | 0.0139 (0.0076) | 0.0085 (0.0074) |
| $\beta_{66}$ | 0.1829 (0.2721) | 0.1259 (0.0317) | 0.2324 (0.0292) |
| $\beta_{1T}$ | 0.0002 (0.0013) | 0.0003 (0.0002) | 0 0 |
| $\beta_{2T}$ | −0.0002 (0.0014) | $3.82 \times 10^{-5}$ (0.0003) | 0 0 |
| $\beta_{3T}$ | 0.0072 (0.0127) | 0.0053 (0.0004) | 0 0 |
| $\beta_{4T}$ | −0.0037 (0.0043) | −0.0017 (0.0002) | 0 0 |
| $\beta_{5T}$ | −0.0005 (0.0008) | −0.0001 (0.0001) | 0 0 |
| $\beta_{6T}$ | −0.0091 (0.0102) | −0.0038 ($5.87 \times 10^{-4}$) | 0 0 |
| **Equation** | **RSQ** | **RSQ** | **RSQ** |
| 1 | 0.8908 | 0.8858 | 0.8788 |
| 2 | 0.9091 | 0.9101 | 0.9053 |
| 3 | 0.9387 | 0.9330 | 0.8660 |
| 4 | 0.9644 | 0.9637 | 0.9491 |
| 5 | 0.7823 | 0.7903 | 0.7833 |

| (4) Homothetic, nonstationary, energy separability, 1960-1974 | (5) Homothetic, nonstationary, with additivity, 1960-1974 | (6) Homothetic, nonstationary, data at four-year intervals, 1962-1973 |
|---|---|---|
| −0.0271 (0.0104) | 0 | −0.0220 (0.0109) |
| −0.0009 (0.0094) | 0 | −0.0217 (0.0133) |
| 0.0168 (0.0134) | 0 | 0.0135 (0.0135) |
| 0 | 0 | 0.0052 (0.0065) |
| 0.0307 (0.0101) | 0 | 0.0440 (0.0121) |
| 0 | 0 | 0.0090 (0.0060) |
| 0 | 0 | 0.0082 (0.0070) |
| 0 | 0 | −0.0149 (0.0088) |
| 0 | 0 | 0.0052 (0.0065) |
| 0 | 0 | 0.0066 (0.0063) |
| 0 | 0 | 0.0023 (0.0078) |
| −0.0498 (0.0090) | 0 | −0.0686 (0.0094) |
| −0.0531 (0.0138) | 0 | −0.0677 (0.0151) |
| −0.0855 (0.0202) | 0 | −0.0867 (0.0256) |
| 0.0307 (0.0101) | 0 | 0.0440 (0.0121) |
| 0 | 0 | 0.0023 (0.0078) |
| 0.1577 (0.0297) | 0 | 0.1767 (0.0360) |
| 0.0004 (0.0002) | 0.0013 (0.0001) | 0.0006 (0.0002) |
| $8.53 \times 10^{-5}$ (0.0003) | 0.0003 (0.0002) | $-3.22 \times 10^{-5}$ (0.0003) |
| 0.0055 (0.0004) | 0.0062 (0.0003) | 0.0056 (0.0004) |
| −0.0019 (0.0002) | −0.0020 (0.0001) | −0.0018 (0.0002) |
| −0.0004 (0.0001) | −0.0040 (0.0001) | −0.0003 (0.0001) |
| −0.0036 ($5.91 \times 10^{-4}$) | −0.0053 ($4.88 \times 10^{-4}$) | −0.0042 (0.0005) |
| **RSQ** | **RSQ** | **RSQ** |
| 0.8848 | 0.8503 | 0.9114 |
| 0.9092 | 0.8864 | 0.9252 |
| 0.9298 | 0.9224 | 0.9391 |
| 0.9632 | 0.9581 | 0.9646 |
| 0.7775 | 0.7775 | 0.8447 |

**Table 4.4**
Price elasticities for "preferred" consumption model

| Elasticity[a] | Year | Belgium | Canada | France | Italy |
|---|---|---|---|---|---|
| $\eta_{AA}$ | 1965 | −1.57 | −1.59 | −1.47 | −1.54 |
| | | (0.12) | (0.13) | (0.10) | (0.12) |
| | 1973 | −1.67 | −1.57 | −1.6 | −1.54 |
| | | (0.15) | (0.12) | (0.13) | (0.12) |
| $\eta_{DD}$ | 1965 | −1.43 | −1.70 | −1.61 | −1.71 |
| | | (0.12) | (0.19) | (0.16) | (0.19) |
| | 1973 | −1.35 | −1.60 | −1.63 | −1.88 |
| | | (0.09) | (0.16) | (0.17) | (0.24) |
| $\eta_{FF}$ | 1965 | −1.04 | −1.05 | −1.03 | −1.03 |
| | | (0.06) | (0.07) | (0.05) | (0.04) |
| | 1973 | −1.04 | −1.06 | −1.05 | −1.04 |
| | | (0.06) | (0.08) | (0.06) | (0.05) |
| $\eta_{TT}$ | 1965 | −1.16 | −1.10 | −1.17 | −1.18 |
| | | (0.13) | (0.08) | (0.14) | (0.15) |
| | 1973 | −1.14 | −1.10 | −1.14 | −1.14 |
| | | (0.12) | (0.09) | (0.12) | (0.12) |
| $\eta_{EE}$ | 1965 | −1.08 | −1.13 | −1.12 | −1.13 |
| | | (0.11) | (0.18) | (0.18) | (0.20) |
| | 1973 | −1.06 | −1.15 | −1.11 | −1.12 |
| | | (0.10) | (0.22) | (0.16) | (0.18) |
| $\eta_{RR}$ | 1965 | −1.37 | −1.30 | −1.39 | −1.45 |
| | | (0.09) | (0.07) | (0.09) | (0.11) |
| | 1973 | −1.40 | −1.30 | −1.30 | −1.32 |
| | | (0.10) | (0.07) | (0.07) | (0.08) |
| $\eta_{EA}$ | 1965 | 0.08 | 0.14 | 0.13 | 0.14 |
| | | (0.10) | (0.17) | (0.17) | (0.18) |
| | 1973 | 0.07 | 0.16 | 0.11 | 0.13 |
| | | (0.09) | (0.20) | (0.14) | (0.17) |
| $\eta_{ED}$ | 1965 | 0.08 | 0.12 | 0.12 | 0.13 |
| | | (0.13) | (0.21) | (0.20) | (0.22) |
| | 1973 | 0.06 | 0.15 | 0.10 | 0.12 |
| | | (0.11) | (0.25) | (0.18) | (0.21) |
| $\eta_{EF}$ | 1965 | 0.43 | 0.70 | 0.68 | 0.74 |
| | | (0.14) | (0.22) | (0.21) | (0.23) |
| | 1973 | 0.37 | 0.82 | 0.59 | 0.68 |
| | | (0.12) | (0.26) | (0.19) | (0.21) |

[a] *A* for apparel; *D*, durables; *F*, food; *T*, transportation and communication; *E*, energy; *R*, all other.

| Netherlands | Norway | U.K. | U.S. | West Germany |
|---|---|---|---|---|
| −1.42 | −1.36 | −1.54 | −1.60 | −1.40 |
| (0.09) | (0.08) | (0.12) | (0.13) | (0.09) |
| −1.51 | −1.52 | −1.57 | −1.62 | −1.45 |
| (0.11) | (0.11) | (0.13) | (0.14) | (0.10) |
| −1.34 | −1.49 | −1.68 | −1.35 | −1.39 |
| (0.09) | (0.13) | (0.18) | (0.09) | (0.10) |
| −1.48 | −1.65 | −1.70 | −1.35 | −1.40 |
| (0.13) | (0.17) | (0.19) | (0.09) | (0.11) |
| −1.04 | −1.03 | −1.04 | −1.05 | −1.04 |
| (0.05) | (0.04) | (0.05) | (0.07) | (0.06) |
| −1.05 | −1.04 | −1.04 | −1.06 | −1.05 |
| (0.06) | (0.06) | (0.05) | (0.08) | (0.06) |
| −1.30 | −1.15 | −1.14 | −1.55 | −1.15 |
| (0.25) | (0.12) | (0.12) | (0.46) | (0.12) |
| −1.18 | −1.13 | −1.11 | −1.46 | −1.13 |
| (0.15) | (0.11) | (0.09) | (0.38) | (0.11) |
| −1.09 | −1.13 | −1.08 | −1.11 | −1.09 |
| (0.13) | (0.20) | (0.12) | (0.16) | (0.14) |
| −1.07 | −1.11 | −1.09 | −1.10 | −1.05 |
| (0.10) | (0.17) | (0.13) | (0.15) | (0.08) |
| −1.40 | −1.51 | −1.39 | −1.27 | −1.44 |
| (0.10) | (0.12) | (0.09) | (0.07) | (0.11) |
| −1.33 | −1.34 | −1.36 | −1.26 | −1.44 |
| (0.08) | (0.08) | (0.09) | (0.06) | (0.11) |
| 0.09 | 0.15 | 0.09 | 0.12 | 0.10 |
| (0.12) | (0.18) | (0.11) | (0.15) | (0.13) |
| 0.08 | 0.12 | 0.09 | 0.11 | 0.06 |
| (0.10) | (0.16) | (0.12) | (0.14) | (0.07) |
| 0.09 | 0.13 | 0.08 | 0.11 | 0.09 |
| (0.15) | (0.23) | (0.14) | (0.18) | (0.16) |
| 0.07 | 0.11 | 0.09 | 0.10 | 0.05 |
| (0.12) | (0.19) | (0.15) | (0.17) | (0.09) |
| 0.49 | 0.75 | 0.46 | 0.60 | 0.53 |
| (0.15) | (0.24) | (0.15) | (0.19) | (0.17) |
| 0.40 | 0.64 | 0.49 | 0.53 | 0.31 |
| (0.13) | (0.20) | (0.16) | (0.18) | (0.09) |

**Table 4.4**
(continued)

| Elasticity[a] | Year | Belgium | Canada | France | Italy |
|---|---|---|---|---|---|
| $\eta_{ET}$ | 1965 | −0.25 | −0.40 | −0.39 | −0.42 |
| | | (0.12) | (0.20) | (0.19) | (0.21) |
| | 1973 | −0.21 | −0.47 | −0.34 | −0.39 |
| | | (0.10) | (0.23) | (0.17) | (0.19) |
| $\eta_{ER}$ | 1965 | −0.27 | −0.43 | −0.42 | −0.46 |
| | | (0.14) | (0.23) | (0.23) | (0.25) |
| | 1973 | −0.23 | −0.51 | −0.37 | −0.42 |
| | | (0.12) | (0.28) | (0.20) | (0.23) |
| $\eta_{AE}$ | 1965 | 0.05 | 0.05 | 0.04 | 0.05 |
| | | (0.06) | (0.07) | (0.05) | (0.06) |
| | 1973 | 0.06 | 0.05 | 0.05 | 0.05 |
| | | (0.08) | (0.06) | (0.07) | (0.06) |
| $\eta_{DE}$ | 1965 | 0.03 | 0.05 | 0.05 | 0.05 |
| | | (0.06) | (0.09) | (0.08) | (0.09) |
| | 1973 | 0.03 | 0.05 | 0.05 | 0.07 |
| | | (0.04) | (0.08) | (0.08) | (0.11) |
| $\eta_{FE}$ | 1965 | 0.08 | 0.09 | 0.06 | 0.05 |
| | | (0.02) | (0.03) | (0.02) | (0.02) |
| | 1973 | 0.08 | 0.10 | 0.08 | 0.07 |
| | | (0.02) | (0.03) | (0.03) | (0.02) |
| $\eta_{TE}$ | 1965 | −0.13 | −0.08 | −0.13 | −0.14 |
| | | (0.06) | (0.04) | (0.06) | (0.07) |
| | 1973 | −0.11 | −0.08 | −0.11 | −0.11 |
| | | (0.05) | (0.04) | (0.05) | (0.05) |
| $\eta_{RE}$ | 1965 | −0.04 | −0.03 | −0.04 | −0.05 |
| | | (0.02) | (0.02) | (0.02) | (0.03) |
| | 1973 | −0.05 | −0.03 | −0.03 | −0.04 |
| | | (0.02) | (0.02) | (0.02) | (0.02) |

| Netherlands | Norway | U.K. | U.S. | West Germany |
|---|---|---|---|---|
| −0.28 | −0.43 | −0.26 | −0.34 | −0.30 |
| (0.14) | (0.21) | (0.13) | (0.17) | (0.15) |
| −0.23 | −0.36 | −0.28 | −0.32 | −0.17 |
| (0.11) | (0.18) | (0.14) | (0.16) | (0.08) |
| −0.30 | −0.46 | −0.29 | −0.37 | −0.33 |
| (0.16) | (0.25) | (0.15) | (0.20) | (0.18) |
| −0.25 | −0.40 | −0.31 | −0.35 | −0.19 |
| (0.13) | (0.21) | (0.17) | (0.19) | (0.10) |
| 0.04 | 0.03 | 0.05 | 0.05 | 0.04 |
| (0.05) | (0.04) | (0.06) | (0.07) | (0.05) |
| 0.05 | 0.05 | 0.05 | 0.06 | 0.04 |
| (0.06) | (0.06) | (0.06) | (0.07) | (0.05) |
| 0.03 | 0.04 | 0.05 | 0.03 | 0.03 |
| (0.04) | (0.06) | (0.09) | (0.04) | (0.05) |
| 0.04 | 0.05 | 0.05 | 0.03 | 0.03 |
| (0.06) | (0.08) | (0.09) | (0.04) | (0.05) |
| 0.07 | 0.06 | 0.06 | 0.09 | 0.07 |
| (0.02) | (0.02) | (0.02) | (0.03) | (0.02) |
| 0.08 | 0.07 | 0.07 | 0.11 | 0.08 |
| (0.03) | (0.02) | (0.02) | (0.03) | (0.03) |
| −0.24 | −0.12 | −0.11 | −0.44 | −0.12 |
| (0.12) | (0.06) | (0.05) | (0.22) | (0.06) |
| −0.14 | −0.10 | −0.09 | −0.37 | −0.10 |
| (0.07) | (0.05) | (0.04) | (0.18) | (0.05) |
| −0.05 | −0.06 | −0.04 | −0.03 | −0.05 |
| (0.02) | (0.03) | (0.02) | (0.02) | (0.03) |
| −0.04 | −0.04 | −0.04 | −0.03 | −0.05 |
| (0.02) | (0.02) | (0.02) | (0.02) | (0.03) |

**Table 4.5**
Parameter estimates of fuel expenditure models

| Parameter[a] | (1) 1960-1974, stationary, regionally homogeneous | (2) 1960-1974, regionally homogeneous | (3) 1960-1974 | (4) 1960-1973 |
|---|---|---|---|---|
| $\alpha_1$ | 0.0414 (0.0383) | −0.2151 (0.0761) | | |
| $\alpha_1 D_1$ | | | −0.8124 (0.0644) | −0.8188 (0.1010) |
| $\alpha_1 D_2$ | | | −0.5234 (0.0475) | −0.5491 (0.0739) |
| $\alpha_1 D_3$ | | | −0.6746 (0.0610) | −0.6813 (0.0919) |
| $\alpha_1 D_4$ | | | −0.4964 (0.0647) | −0.4830 (0.1021) |
| $\alpha_1 D_5$ | | | −0.6089 (0.0551) | −0.6181 (0.0843) |
| $\alpha_1 D_6$ | | | −0.3892 (0.0442) | −0.3608 (0.0576) |
| $\alpha_1 D_7$ | | | −0.6954 (0.0569) | −0.6902 (0.0904) |
| $\alpha_1 D_8$ | | | −0.5752 (0.0544) | −0.6261 (0.0848) |
| $\alpha_1 D_9$ | | | −0.4455 (0.0630) | −0.4337 (0.0949) |
| $\alpha_2$ | −0.2360 (0.0228) | −0.0894 (0.0449) | | |
| $\alpha_2 D_1$ | | | −0.1246 (0.0524) | −0.1554 (0.0681) |
| $\alpha_2 D_2$ | | | −0.3949 (0.0363) | −0.4219 (0.0491) |
| $\alpha_2 D_3$ | | | −0.1008 (0.0531) | −0.1293 (0.0665) |
| $\alpha_2 D_4$ | | | −0.0791 (0.0536) | −0.1122 (0.0688) |
| $\alpha_2 D_5$ | | | −0.1994 (0.0475) | −0.2383 (0.0610) |
| $\alpha_2 D_6$ | | | 0.0165 (0.0395) | −0.0030 (0.0430) |
| $\alpha_2 D_7$ | | | 0.0020 (0.0402) | −0.0359 (0.0530) |
| $\alpha_2 D_8$ | | | −0.4154 (0.0406) | −0.4519 (0.0575) |
| $\alpha_2 D_9$ | | | −0.1189 (0.0547) | −0.1569 (0.0680) |
| $\alpha_3$ | −0.0870 (0.0236) | 0.0855 (0.0501) | | |
| $\alpha_3 D_1$ | | | −0.6495 (0.0565) | −0.6474 (0.0684) |
| $\alpha_3 D_2$ | | | −0.3221 (0.0428) | −0.2619 (0.0560) |
| $\alpha_3 D_3$ | | | −0.6950 (0.0557) | −0.6921 (0.0649) |
| $\alpha_3 D_4$ | | | −0.7579 (0.0579) | −0.7777 (0.0693) |
| $\alpha_3 D_5$ | | | −0.6868 (0.0494) | −0.6617 (0.0598) |
| $\alpha_3 D_6$ | | | −0.2238 (0.0452) | −0.2434 (0.0463) |
| $\alpha_3 D_7$ | | | −0.5659 (0.0474) | −0.5816 (0.0571) |
| $\alpha_3 D_8$ | | | −0.3897 (0.0513) | −0.3029 (0.0692) |
| $\alpha_3 D_9$ | | | −0.6287 (0.0586) | −0.6329 (0.0671) |
| $\alpha_4$ | −0.6984 (0.0414) | −0.7810 (0.0832) | | |
| $\alpha_4 D_1$ | | | 0.5866 (0.0695) | 0.6216 (0.1142) |
| $\alpha_4 D_2$ | | | 0.2403 (0.0550) | 0.2328 (0.0874) |
| $\alpha_4 D_3$ | | | 0.4704 (0.0594) | 0.5026 (0.1093) |
| $\alpha_4 D_4$ | | | 0.3334 (0.0683) | 0.3729 (0.1135) |
| $\alpha_4 D_5$ | | | 0.4953 (0.0594) | 0.5181 (0.0968) |
| $\alpha_4 D_6$ | | | −0.4036 (0.0345) | −0.3929 (0.0649) |
| $\alpha_4 D_7$ | | | 0.2593 (0.0575) | 0.3078 (0.0950) |
| $\alpha_4 D_8$ | | | 0.3803 (0.0647) | 0.3808 (0.1030) |

[a] Fuels are numbered 1 = solid, 2 = liquid, 3 = gas, 4 = electricity. Country dummy variables are numbered alphabetically: 1 = Belgium; 2 = Canada, 3 = France, 4 = Italy, 5 = Netherlands, 6 = Norway, 7 = U.K., 8 = U.S., 9 = West Germany.

| (5)<br>1960–1974<br>(Europe only) | (6)<br>1960–1974<br>(U.S. and Canada only) | (7)<br>1960–1974,<br>stationary | (8)<br>1962–1974,<br>regionally homogeneous four-year intervals | (9)<br>1960–1974,<br>additivity |
|---|---|---|---|---|
| | | | −0.2478 (0.1328) | |
| −1.012 (0.1010) | | −0.0796 (0.0529) | | −0.6811 (0.0251) |
| | −0.1021 (0.0259) | 0.0404 (0.0406) | | −0.3160 (0.0258) |
| −0.8505 (0.0575) | | −0.0094 (0.0486) | | −0.5733 (0.0251) |
| −0.6958 (0.0615) | | 0.2301 (0.0528) | | −0.3860 (0.0251) |
| −0.7716 (0.0527) | | 0.0068 (0.0431) | | −0.4853 (0.0251) |
| −0.5580 (0.0389) | | 0.0810 (0.0310) | | −0.4056 (0.0283) |
| −0.9261 (0.0545) | | −0.0045 (0.0481) | | −0.5974 (0.0251) |
| | −0.1099 (0.0287) | 0.0131 (0.0514) | | −0.2952 (0.0251) |
| −0.6327 (0.0587) | | 0.2368 (0.0502) | | −0.3692 (0.0256) |
| | | | −0.0593 (0.0875) | |
| −0.0533 (0.0553) | −0.5274 (0.0496) | −0.3251 (0.0419) | | −0.0497 (0.0245) |
| | | −0.5518 (0.0312) | | −0.2353 (0.0253) |
| −0.0331 (0.0562) | | −0.2952 (0.0415) | | −0.0484 (0.0245) |
| −0.0031 (0.0566) | | −0.2807 (0.0423) | | −0.0231 (0.0245) |
| −0.1436 (0.0502) | | −0.3815 (0.0367) | | −0.1223 (0.0245) |
| 0.1033 (0.0422) | | −0.1498 (0.0256) | | −0.0247 (0.0277) |
| 0.0892 (0.0418) | | −0.1821 (0.0317) | | 0.0535 (0.0245) |
| | −0.4467 (0.0559) | −0.5703 (0.0398) | | −0.1903 (0.0245) |
| −0.0407 (0.0580) | | −0.3184 (0.0424) | | −0.0911 (0.0250) |
| | | | 0.0968 (0.0941) | |
| −0.6964 (0.0593) | −0.0414 (0.0313) | −0.6066 (0.0448) | | −0.0485 (0.0299) |
| | | −0.2776 (0.0422) | | −0.0278 (0.0308) |
| −0.7414 (0.0587) | | −0.6530 (0.0420) | | −0.1139 (0.0299) |
| −0.8030 (0.0605) | | −0.7171 (0.0445) | | −0.1429 (0.0299) |
| −0.7285 (0.0528) | | −0.6435 (0.0380) | | −0.1871 (0.0299) |
| −0.2069 (0.0475) | | −0.2062 (0.0285) | | 0.1427 (0.0337) |
| −0.5698 (0.0484) | | −0.5536 (0.0355) | | −0.1099 (0.0299) |
| | −0.1053 (0.0372) | −0.3347 (0.0574) | | −0.1092 (0.0299) |
| −0.6704 (0.0612) | | −0.5905 (0.0432) | | −0.0197 (0.0305) |
| | | | −0.7896 (0.1392) | |
| 0.7616 (0.0730) | −0.3291 (0.0304) | 0.0113 (0.0524) | | −0.2207 (0.0238) |
| | | −0.2109 (0.0452) | | −0.4209 (0.0245) |
| 0.6253 (0.0641) | | −0.0423 (0.0450) | | −0.2644 (0.0238) |
| 0.5019 (0.0714) | | −0.2321 (0.0510) | | −0.4481 (0.0238) |
| 0.6437 (0.0578) | | 0.0183 (0.0439) | | −0.2052 (0.0238) |
| −0.3384 (0.0362) | | −0.7250 (0.0233) | | −0.7124 (0.0180) |
| 0.4066 (0.0590) | | −0.2596 (0.0419) | | −0.3463 (0.0238) |
| | −0.3382 (0.0361) | −0.1081 (0.0591) | | −0.4053 (0.0238) |

**Table 4.5**
(continued)

| Parameter[a] | (1) 1960–1974, stationary, regionally homogeneous | (2) 1960–1974, regionally homogeneous | (3) 1960–1974 | (4) 1960–1973 |
|---|---|---|---|---|
| $\alpha_4 D_9$ | | | 0.1932 (0.0636) | 0.2235 (0.1048) |
| $\beta_{11}$ | 0.3708 (0.0423) | 0.2657 (0.0415) | −0.0012 (0.0352) | 0.0304 (0.0516) |
| $\beta_{12}$ | −0.1265 (0.0239) | −0.0807 (0.0223) | 0.0203 (0.0227) | 0.0159 (0.0302) |
| $\beta_{13}$ | −0.0914 (0.0171) | −0.0803 (0.0158) | −0.1363 (0.0258) | −0.1826 (0.0305) |
| $\beta_{14}$ | −0.1522 (0.0273) | −0.1046 (0.0293) | 0.1171 (0.0227) | 0.1362 (0.0397) |
| $\beta_{21}$ | −0.1265 (0.0239) | −0.0807 (0.0223) | 0.0203 (0.0227) | 0.0159 (0.0302) |
| $\beta_{22}$ | 0.0286 (0.0178) | 0.0150 (0.0164) | 0.0162 (0.0252) | 0.0108 (0.0315) |
| $\beta_{23}$ | 0.0917 (0.0109) | 0.0933 (0.0105) | −0.1285 (0.0208) | −0.1298 (0.0245) |
| $\beta_{24}$ | 0.0063 (0.0151) | −0.0276 (0.0155) | 0.0920 (0.0146) | 0.1031 (0.0242) |
| $\beta_{31}$ | −0.0914 (0.0273) | −0.0803 (0.0158) | −0.1363 (0.0258) | −0.1826 (0.0305) |
| $\beta_{32}$ | 0.0917 (0.0109) | 0.0933 (0.0105) | −0.1285 (0.0208) | −0.1298 (0.0245) |
| $\beta_{33}$ | 0.0747 (0.0159) | 0.0836 (0.0147) | 0.1207 (0.0319) | 0.1853 (0.0359) |
| $\beta_{34}$ | −0.0750 (0.0168) | −0.0966 (0.0169) | 0.1440 (0.0193) | 0.1270 (0.0285) |
| $\beta_{41}$ | −0.1522 (0.0273) | −0.1046 (0.0293) | 0.1171 (0.0227) | 0.1362 (0.0397) |
| $\beta_{42}$ | 0.0063 (0.0151) | −0.0276 (0.0155) | 0.0920 (0.0146) | 0.1031 (0.0242) |
| $\beta_{43}$ | −0.0750 (0.0168) | −0.0966 (0.0169) | 0.1440 (0.0193) | 0.1270 (0.0285) |
| $\beta_{44}$ | 0.2209 (0.0294) | 0.2289 (0.0323) | −0.3531 (0.0296) | −0.3663 (0.0484) |
| $\beta_{1T}$ | 0 | 0.0112 (0.0027) | 0.0181 (0.0014) | 0.0179 (0.0018) |
| $\beta_{2T}$ | 0 | −0.0068 (0.0017) | −0.0068 (0.0012) | −0.0060 (0.0013) |
| $\beta_{3T}$ | 0 | −0.0079 (0.0021) | 0.0009 (0.0013) | −0.0005 (0.0014) |
| $\beta_{4T}$ | 0 | 0.0034 (0.0029) | −0.0121 (0.0011) | −0.0114 (0.0017) |
| **RSQ** | | | | |
| Equation 1 | 0.375 | 0.464 | 0.894 | 0.904 |
| Equation 2 | 0.437 | 0.455 | 0.772 | 0.785 |
| Equation 3 | 0.114 | 0.210 | 0.746 | 0.769 |

| (5) 1960-1974 (Europe only) | (6) 1960-1974 (U.S. and Canada only) | (7) 1960-1974, stationary | (8) 1962-1974, regionally homogeneous four-year intervals | (9) 1960-1974, additivity |
|---|---|---|---|---|
| 0.3439 (0.0663) | | −0.3275 (0.0396) | | −0.5199 (0.0243) |
| −0.1199 (0.0300) | 0.0179 (0.0137) | 0.2620 (0.0456) | 0.2153 (0.0753) | 0 |
| 0.0716 (0.0195) | −0.0161 (0.0067) | −0.0089 (0.0290) | −0.0544 (0.0409) | 0 |
| −0.1123 (0.0224) | −0.0353 (0.0064) | −0.1515 (0.0284) | −0.0706 (0.0292) | 0 |
| 0.1606 (0.0236) | 0.0336 (0.0160) | −0.1016 (0.0282) | −0.0903 (0.0497) | 0 |
| 0.0716 (0.0195) | −0.0161 (0.0067) | −0.0089 (0.0290) | −0.0544 (0.0409) | 0 |
| 0.0016 (0.0261) | −0.1665 (0.0236) | 0.0012 (0.0273) | 0.0069 (0.0311) | 0 |
| −0.1698 (0.0229) | 0.1322 (0.0146) | −0.1142 (0.0221) | 0.0843 (0.0203) | 0 |
| 0.0965 (0.0157) | 0.0504 (0.0240) | 0.1212 (0.0203) | −0.0368 (0.0279) | 0 |
| −0.1123 (0.0224) | −0.0353 (0.0064) | 0.1515 (0.0284) | −0.0706 (0.0292) | 0 |
| −0.1698 (0.0229) | 0.1322 (0.0146) | −0.1142 (0.0221) | 0.0843 (0.0203) | 0 |
| 0.1044 (0.0336) | −0.0243 (0.0167) | −0.1383 (0.0348) | 0.0727 (0.0279) | 0 |
| 0.1777 (0.0216) | −0.0726 (0.0158) | 0.1274 (0.0271) | −0.0864 (0.0291) | 0 |
| 0.1606 (0.0236) | 0.0336 (0.0160) | −0.1016 (0.0282) | −0.0903 (0.0497) | 0 |
| 0.0965 (0.0157) | 0.0504 (0.0240) | 0.1219 (0.0203) | −0.0368 (0.0279) | 0 |
| 0.1777 (0.0216) | −0.0726 (0.0158) | 0.1274 (0.0271) | −0.0864 (0.0291) | 0 |
| −0.4348 (0.0313) | −0.0114 (0.0334) | −0.1477 (0.0320) | 0.2134 (0.0523) | 0 |
| 0.0254 (0.0013) | 0.0019 (0.0004) | 0 | 0.0121 (0.0047) | 0.0164 (0.0011) |
| −0.0100 (0.0013) | 0.0060 (0.0011) | 0 | −0.0076 (0.0033) | −0.0075 (0.0011) |
| −0.0009 (0.0015) | −0.0011 (0.0007) | 0 | −0.0092 (0.0039) | −0.0073 (0.0014) |
| −0.0146 (0.0012) | −0.0067 (0.0013) | 0 | 0.0046 (0.0049) | −0.0016 (0.0011) |
| | | | | |
| 0.940 | 0.899 | 0.7911 | 0.4603 | 0.880 |
| 0.667 | 0.821 | 0.7185 | 0.4093 | 0.739 |
| 0.785 | 0.960 | 0.7340 | 0.1889 | 0.636 |

**Table 4.6**
Own-price elasticities for models 2 and 8

A. Model 2: Nonstationary, regionally homogeneous, annual data 1960–1974.

| Elasticity[a] | Year | Belgium | Canada | France | Italy |
|---|---|---|---|---|---|
| $\eta_{11}$ | 1965 | −1.56 | [b] | −1.74 | −3.21 |
| | | (0.08) | | (0.12) | (0.34) |
| | 1973 | −2.48 | [b] | −4.45 | −12.64 |
| | | (0.23) | | (0.54) | (1.81) |
| $\eta_{22}$ | 1965 | −1.12 | −1.04 | −1.13 | −1.11 |
| | | (0.13) | (0.04) | (0.15) | (0.12) |
| | 1973 | −1.04 | −1.04 | −1.04 | −1.06 |
| | | (0.05) | (0.04) | (0.05) | (0.06) |
| $\eta_{33}$ | 1965 | −1.48 | −1.46 | −1.34 | −1.31 |
| | | (0.09) | (0.08) | (0.06) | (0.05) |
| | 1973 | −1.41 | −1.56 | −1.31 | −1.32 |
| | | (0.07) | (0.10) | (0.06) | (0.06) |
| $\eta_{44}$ | 1965 | −1.98 | −1.53 | −1.81 | −1.48 |
| | | (0.13) | (0.08) | (0.11) | (0.07) |
| | 1973 | −1.89 | −1.49 | −1.74 | −1.49 |
| | | (0.13) | (0.07) | (0.11) | (0.07) |

B. Model 8: Nonstationary, regionally homogeneous, data at four-year intervals, 1962–1964

| Elasticity[a] | Year | Belgium | Canada | France | Italy |
|---|---|---|---|---|---|
| $\eta_{11}$ | 1965 | −1.45 | [b] | −1.60 | −2.78 |
| | | (0.16) | | (0.21) | (0.63) |
| | 1973 | −2.20 | [b] | −3.79 | −10.40 |
| | | (0.42) | | (0.97) | (3.27) |
| $\eta_{22}$ | 1965 | −1.06 | −1.02 | −1.06 | −1.05 |
| | | (2.5) | (0.08) | (0.27) | (0.23) |
| | 1973 | −1.02 | −1.02 | −1.02 | −1.03 |
| | | (0.07) | (0.08) | (0.09) | (0.12) |
| $\eta_{33}$ | 1965 | −1.43 | −1.40 | −1.29 | −1.27 |
| | | (0.16) | (.15) | (0.11) | (0.10) |
| | 1973 | −1.36 | −1.49 | −1.27 | −1.28 |
| | | (0.14) | (0.17) | (0.10) | (0.11) |
| $\eta_{44}$ | 1965 | −1.92 | −1.50 | −1.75 | −1.45 |
| | | (0.23) | (0.12) | (0.19) | (0.11) |
| | 1973 | −1.83 | −1.46 | −1.70 | −1.45 |
| | | (0.20) | (0.11) | (0.17) | (0.11) |

[a] 1 = solid, 2 = liquid, 3 = gas, 4 = electricity.
[b] Almost no solid fuel is consumed in the residential sectors of Canada and the U.S., and almost no natural gas is consumed in Norway, so that these elasticities are meaningless.

| Netherlands | Norway | U.K. | U.S. | West Germany |
|---|---|---|---|---|
| −1.94 | −3.71 | −1.80 | b | −3.87 |
| (0.15) | (0.42) | (0.13) | | (0.49) |
| −15.07 | −10.58 | −1.83 | b | −13.67 |
| (2.19) | (1.50) | (0.21) | | (2.00) |
| −1.05 | −1.12 | −1.23 | −1.04 | −1.07 |
| (0.06) | (0.13) | (0.26) | (0.04) | (0.08) |
| −1.08 | −1.08 | −1.27 | −1.04 | −1.04 |
| (0.08) | (0.08) | (0.29) | (0.04) | (0.05) |
| −1.49 | b | −1.44 | −1.33 | −1.69 |
| (0.08) | | (0.07) | (0.06) | (0.12) |
| −1.14 | b | −1.27 | −1.46 | −1.40 |
| (0.02) | | (0.05) | (0.08) | (0.07) |
| −1.86 | −1.30 | −1.54 | −1.56 | −1.40 |
| (0.12) | (0.04) | (0.07) | (0.08) | (0.06) |
| −2.21 | −1.30 | −1.52 | −1.52 | −1.51 |
| (0.17) | (0.04) | (0.07) | (0.07) | (0.07) |

| Netherlands | Norway | U.K. | U.S. | West Germany |
|---|---|---|---|---|
| −1.76 | −3.19 | −1.66 | b | −3.33 |
| (0.27) | (0.76) | (0.23) | | (0.81) |
| −12.41 | −8.76 | −2.11 | b | −11.27 |
| (3.94) | (2.69) | (0.39) | | (3.55) |
| −1.02 | −1.05 | −1.11 | −1.02 | −1.03 |
| (0.11) | (0.25) | (0.50) | (0.09) | (0.15) |
| −1.03 | −1.03 | −1.12 | −1.02 | −1.02 |
| (0.16) | (0.16) | (0.55) | (0.08) | (0.09) |
| −1.43 | b | −1.38 | −1.28 | −1.60 |
| (0.16) | | (0.15) | (0.10) | (0.23) |
| −1.12 | b | −1.23 | −1.40 | −1.35 |
| (0.05) | | (0.09) | (0.15) | (0.13) |
| −1.80 | −1.27 | −1.51 | −1.52 | −1.37 |
| (0.20) | (0.07) | (0.12) | (0.13) | (0.09) |
| −2.12 | −1.28 | −1.49 | −1.48 | −1.48 |
| (0.28) | (0.07) | (0.12) | (0.12) | (0.12) |

**Table 4.7**
Partial price elasticities, U.S. and Canada estimated separately from Europe

| Elasticity[a] | Year | Belgium | Canada | France | Italy |
|---|---|---|---|---|---|
| $\eta_{11}$ | 1965 | −0.75 | [b] | −0.66 | −0.00 |
| | | (0.06) | | (0.08) | (0.25) |
| | 1973 | −0.33 | [b] | 0.56 | 4.26 |
| | | (0.17) | | (0.39) | (1.31) |
| $\eta_{12}$ | 1965 | −0.15 | [b] | −0.20 | −0.59 |
| | | (0.04) | | (0.05) | (0.16) |
| | 1973 | −0.40 | [b] | −0.93 | −3.14 |
| | | (0.12) | | (0.24) | (0.84) |
| $\eta_{13}$ | 1965 | 0.24 | [b] | 0.31 | 0.93 |
| | | (0.05) | | (0.06) | (0.18) |
| | 1973 | 0.63 | [b] | 1.46 | 4.92 |
| | | (0.12) | | (0.29) | (0.98) |
| $\eta_{14}$ | 1965 | −0.34 | [b] | −0.45 | −1.33 |
| | | (0.05) | | (0.06) | (0.19) |
| | 1973 | −0.90 | [b] | −2.08 | −7.04 |
| | | (0.13) | | (0.30) | (1.03) |
| $\eta_{21}$ | 1965 | −0.58 | 0.04 | −0.63 | −0.53 |
| | | (0.16) | (0.02) | (0.17) | (0.14) |
| | 1973 | −0.20 | 0.04 | −0.20 | −0.28 |
| | | (0.05) | (0.02) | (0.05) | (0.08) |
| $\eta_{22}$ | 1965 | −1.01 | −0.55 | −1.01 | −1.01 |
| | | (0.21) | (0.06) | (0.23) | (0.19) |
| | 1973 | −1.00 | −0.56 | −1.00 | −1.01 |
| | | (0.07) | (0.06) | (0.07) | (0.10) |
| $\eta_{23}$ | 1965 | 1.38 | −0.35 | 1.50 | 1.25 |
| | | (0.18) | (0.04) | (0.20) | (0.17) |
| | 1973 | 0.47 | −0.35 | 0.48 | 0.66 |
| | | (0.06) | (0.04) | (0.06) | (0.09) |
| $\eta_{24}$ | 1965 | −0.78 | −0.13 | −0.85 | −0.71 |
| | | (0.13) | (0.06) | (0.14) | (0.11) |
| | 1973 | −0.27 | −0.13 | −0.27 | −0.38 |
| | | (0.04) | (0.06) | (0.04) | (0.06) |
| $\eta_{31}$ | 1965 | 0.66 | 0.19 | 0.45 | 0.42 |
| | | (0.13) | (0.03) | (0.09) | (0.08) |
| | 1973 | 0.55 | 0.24 | 0.42 | 0.44 |
| | | (0.11) | (0.04) | (0.08) | (0.09) |

[a] 1 = solid, 2 = liquid, 3 = gas, 4 = electricity.
[b] Almost no solid fuel is consumed in the residential sectors of Canada and the U.S., and almost no natural gas is consumed in Norway, so that these elasticities are meaningless.

| Netherlands | Norway | U.K. | U.S. | West Germany |
|---|---|---|---|---|
| -0.57 | 0.22 | -0.63 | [b] | 0.32 |
| (0.11) | (0.30) | (0.09) | | (0.32) |
| 5.35 | 3.32 | -0.38 | [b] | 4.72 |
| (1.59) | (1.08) | (0.15) | | (1.43) |
| -0.25 | -0.73 | -0.22 | [b] | -0.78 |
| (0.07) | (0.20) | (0.06) | | (0.21) |
| -3.79 | -2.58 | -0.37 | [b] | -3.42 |
| (0.95) | (0.41) | (0.05) | | (0.73) |
| 0.40 | 1.14 | 0.34 | [b] | 1.22 |
| (0.08) | (0.23) | (0.07) | | (0.24) |
| 5.95 | 4.05 | 0.58 | [b] | 5.36 |
| (1.18) | (0.81) | (0.11) | | (1.07) |
| -0.57 | -1.63 | -0.49 | [b] | -1.74 |
| (0.08) | (0.24) | (0.07) | | (0.25) |
| -8.51 | -5.79 | -0.83 | [b] | -7.66 |
| (1.25) | (0.85) | (0.12) | | (1.12) |
| -0.25 | -0.57 | -1.14 | 0.048 | -0.33 |
| (0.07) | (0.15) | (0.31) | (0.02) | (0.09) |
| -0.37 | -0.36 | -1.28 | 0.04 | -0.22 |
| (0.10) | (0.10) | (0.35) | (0.021) | (0.06) |
| -1.00 | -1.01 | -1.02 | -0.50 | -1.01 |
| (0.09) | (0.21) | (0.42) | (0.07) | (0.12) |
| -1.01 | -1.01 | -1.03 | -0.56 | -1.00 |
| (0.13) | (0.13) | (0.46) | (0.06) | (0.08) |
| -0.60 | 1.36 | 2.71 | -0.39 | 0.79 |
| (0.08) | (0.18) | (0.36) | (0.04) | (0.11) |
| 0.87 | 0.85 | 3.03 | -0.35 | 0.52 |
| (0.12) | (0.11) | (0.41) | (0.04) | (0.07) |
| -0.34 | -0.77 | -1.54 | -0.15 | -0.45 |
| (0.05) | (0.12) | (0.25) | (0.07) | (0.07) |
| -0.49 | -0.48 | -1.72 | -0.13 | -0.29 |
| (0.08) | (0.08) | (0.28) | (0.06) | (0.05) |
| 0.66 | [b] | 0.59 | 0.14 | 0.93 |
| (0.13) | | (0.12) | (0.03) | (0.18) |
| 0.19 | [b] | 0.36 | 0.19 | 0.54 |
| (0.04) | | (0.07) | (0.03) | (0.11) |

**Table 4.7**
(continued)

| Elasticity[a] | Year | Belgium | Canada | France | Italy |
|---|---|---|---|---|---|
| $\eta_{32}$ | 1965 | 0.99 | −0.73 | 0.68 | 0.63 |
| | | (0.13) | (0.08) | (0.09) | (0.08) |
| | 1973 | 0.83 | −0.88 | 0.64 | 0.66 |
| | | (0.11) | (0.10) | (0.08) | (0.09) |
| $\eta_{33}$ | 1965 | −1.61 | −0.87 | −1.42 | −1.39 |
| | | (0.20) | (0.09) | (0.13) | (0.12) |
| | 1973 | −1.51 | −0.84 | −1.39 | −1.41 |
| | | (0.16) | (0.11) | (0.13) | (0.13) |
| $\eta_{34}$ | 1965 | −1.04 | 0.40 | −0.71 | −0.66 |
| | | (0.13) | (0.09) | (0.09) | (0.08) |
| | 1973 | −0.87 | 0.48 | −0.67 | −0.69 |
| | | (0.10) | (0.10) | (0.08) | (0.08) |
| $\eta_{41}$ | 1965 | −0.69 | −0.08 | −0.57 | −0.34 |
| | | (0.10) | (0.04) | (0.08) | (0.05) |
| | 1973 | −0.63 | −0.07 | −0.52 | −0.34 |
| | | (0.09) | (0.03) | (0.08) | (0.05) |
| $\eta_{42}$ | 1965 | −0.41 | −0.12 | −0.34 | −0.20 |
| | | (0.07) | (0.06) | (0.05) | (0.03) |
| | 1973 | −0.38 | −0.11 | −0.32 | −0.21 |
| | | (0.06) | (0.05) | (0.51) | (0.03) |
| $\eta_{43}$ | 1965 | −0.76 | 0.17 | −0.63 | −0.37 |
| | | (0.09) | (0.04) | (0.08) | (0.04) |
| | 1973 | −0.69 | 0.15 | −0.58 | −0.38 |
| | | (0.08) | (0.03) | (0.07) | (0.05) |
| $\eta_{44}$ | 1965 | 0.87 | −0.97 | 0.55 | −0.09 |
| | | (0.13) | (0.08) | (0.11) | (0.06) |
| | 1973 | 0.70 | −0.97 | 0.42 | −0.06 |
| | | (0.12) | (0.07) | (0.10) | (0.07) |

| Netherlands | Norway | U.K. | U.S. | West Germany |
|---|---|---|---|---|
| 1.00 | b | 0.89 | −0.52 | 1.40 |
| (0.13) | | (0.12) | (0.06) | (0.19) |
| 0.28 | b | 0.54 | −0.73 | 0.82 |
| (0.04) | | (0.07) | (0.08) | (0.11) |
| −1.61 | b | −1.55 | −0.90 | −1.86 |
| (0.20) | | (0.18) | (0.06) | (0.28) |
| −1.17 | b | −1.33 | −0.87 | −1.50 |
| (0.06) | | (0.11) | (0.09) | (0.16) |
| −1.05 | b | −0.94 | 0.29 | −1.47 |
| (0.13) | | (0.11) | (0.06) | (0.18) |
| −0.30 | b | −0.57 | 0.40 | −0.86 |
| (0.04) | | (0.07) | (0.09) | (0.10) |
| −0.60 | −0.21 | −0.38 | −0.08 | −0.28 |
| (0.09) | (0.03) | (0.06) | (0.04) | (0.04) |
| −0.85 | −0.21 | −0.37 | −0.08 | −0.36 |
| (0.12) | (0.03) | (0.05) | (0.04) | (0.05) |
| −0.36 | −0.12 | −0.23 | −0.12 | −0.17 |
| (0.06) | (0.02) | (0.04) | (0.06) | (0.03) |
| −0.51 | −0.12 | −0.22 | −0.11 | −0.22 |
| (0.08) | (0.02) | (0.03) | (0.05) | (0.03) |
| −0.67 | −0.23 | −0.42 | 0.18 | −0.31 |
| (0.08) | (0.028) | (0.05) | (0.04) | (0.04) |
| −0.94 | −0.23 | −0.40 | 0.16 | −0.40 |
| (0.11) | (0.028) | (0.05) | (0.03) | (0.05) |
| 0.63 | −0.44 | 0.04 | −0.97 | −0.24 |
| (0.12) | (0.04) | (0.07) | (0.08) | (0.05) |
| 1.29 | −0.43 | −0.01 | −0.97 | −0.02 |
| (0.16) | (0.04) | (0.07) | (0.07) | (0.07) |

**Table 4.8**
Partial price elasticities for model 3, all nine countries pooled together

| Elasticity[a] | Year | Belgium | Canada | France | Italy |
|---|---|---|---|---|---|
| $\eta_{11}$ | 1965 | −0.99 | [b] | −0.99 | −0.99 |
| | | (0.07) | | (0.09) | (0.29) |
| | 1973 | −0.99 | [b] | −0.98 | −0.95 |
| | | (0.19) | | (0.46) | (1.50) |
| $\eta_{12}$ | 1965 | −0.04 | [b] | −0.06 | −0.17 |
| | | (0.05) | | (0.06) | (0.19) |
| | 1973 | −0.11 | [b] | −0.26 | −0.89 |
| | | (0.13) | | (0.29) | (1.00) |
| $\eta_{13}$ | 1965 | 0.29 | [b] | 0.38 | 1.13 |
| | | (0.05) | | (0.07) | (0.21) |
| | 1973 | 0.76 | [b] | 1.77 | 5.97 |
| | | (0.14) | | (0.33) | (1.13) |
| $\eta_{14}$ | 1965 | −0.25 | [b] | −0.33 | −0.97 |
| | | (0.05) | | (0.06) | (0.19) |
| | 1973 | −0.65 | [b] | −1.52 | −5.13 |
| | | (0.13) | | (0.29) | (1.0) |
| $\eta_{21}$ | 1965 | −0.16 | −0.05 | −0.18 | −0.15 |
| | | (0.18) | (0.06) | (0.20) | (0.17) |
| | 1973 | −0.06 | −0.05 | −0.06 | −0.08 |
| | | (0.06) | (0.06) | (0.06) | (0.09) |
| $\eta_{22}$ | 1965 | −1.13 | −1.04 | −1.14 | −1.11 |
| | | (0.20) | (0.07) | (0.22) | (0.09) |
| | 1973 | −1.04 | −1.04 | −1.04 | −1.06 |
| | | (0.07) | (0.07) | (0.07) | (0.10) |
| $\eta_{23}$ | 1965 | 1.04 | 0.34 | 1.14 | 0.95 |
| | | (0.17) | (0.05) | (0.18) | (0.15) |
| | 1973 | 0.36 | 0.34 | 0.37 | 0.50 |
| | | (0.06) | (0.05) | (0.06) | (0.08) |
| $\eta_{24}$ | 1965 | −0.75 | −0.24 | −0.81 | −0.68 |
| | | (0.11) | (0.04) | (0.13) | (0.11) |
| | 1973 | −0.25 | −0.24 | −0.26 | −0.36 |
| | | (0.04) | (0.04) | (0.04) | (0.06) |
| $\eta_{31}$ | 1965 | 0.80 | 0.75 | 0.55 | 0.51 |
| | | (0.15) | (0.14) | (0.10) | (0.10) |
| | 1973 | 0.67 | 0.91 | 0.51 | 0.53 |
| | | (0.13) | (0.17) | (0.10) | (0.10) |

[a] 1 = solid, 2 = liquid, 3 = gas, 4 = electricity.
[b] Almost no solid fuel is consumed in the residential sectors of Canada and the U.S., and almost no natural gas is consumed in Norway, so that these elasticities are meaningless.

| Netherlands | Norway | U.K. | U.S. | West Germany |
|---|---|---|---|---|
| −0.99 | −0.99 | −0.99 | [b] | −0.99 |
| (0.12) | (0.36) | (0.11) | | (0.38) |
| −0.94 | −0.96 | −0.99 | [b] | −0.94 |
| (1.86) | (1.27) | (0.18) | | (1.68) |
| −0.07 | −0.21 | −0.06 | [b] | −0.22 |
| (0.08) | (0.23) | (0.07) | | (0.25) |
| −1.07 | −0.73 | −0.10 | [b] | −0.97 |
| (1.20) | (0.82) | (0.12) | | (1.08) |
| 0.48 | 1.39 | 0.41 | [b] | 1.47 |
| (0.09) | (0.26) | (0.08) | | (.28) |
| 7.22 | 4.91 | 0.71 | [b] | 6.50 |
| (1.37) | (0.93) | (0.13) | | (1.23) |
| −0.42 | −1.19 | −0.36 | [b] | −1.27 |
| (0.08) | (0.23) | (0.07) | | (0.25) |
| −6.20 | −4.22 | −0.61 | [b] | −5.59 |
| (1.2) | (0.82) | (0.12) | | (1.08) |
| −0.07 | −0.16 | −0.32 | −0.06 | −0.09 |
| (0.08) | (0.18) | (0.36) | (0.07) | (0.11) |
| −0.10 | −0.10 | −0.36 | −0.05 | −0.06 |
| (0.12) | (0.11) | (0.40) | (0.06) | (0.07) |
| −1.06 | −1.13 | −1.26 | −1.05 | −1.07 |
| (0.09) | (0.20) | (0.40) | (0.07) | (0.12) |
| −1.08 | −1.08 | −1.29 | −1.04 | −1.05 |
| (0.13) | (0.13) | (0.45) | (0.07) | (0.08) |
| 0.45 | 1.03 | 2.05 | 0.38 | 0.60 |
| (0.07) | (0.17) | (0.33) | (0.06) | (0.10) |
| 0.66 | 0.64 | 2.29 | 0.34 | 0.39 |
| (0.11) | (0.10) | (0.37) | (0.05) | (0.06) |
| −0.32 | −0.07 | −1.47 | −0.27 | −0.43 |
| (0.05) | (0.12) | (0.23) | (0.04) | (0.07) |
| −0.47 | −0.46 | −1.64 | −0.24 | −0.28 |
| (0.07) | (0.07) | (0.26) | (0.04) | (0.04) |
| 0.80 | [b] | 0.72 | 0.54 | 1.13 |
| (0.15) | | (0.14) | (0.10) | (0.21) |
| 0.23 | [b] | 0.43 | 0.75 | 0.66 |
| (0.04) | | (0.08) | (0.14) | (0.12) |

**Table 4.8**
(continued)

| Elasticity[a] | Year | Belgium | Canada | France | Italy |
|---|---|---|---|---|---|
| $\eta_{32}$ | 1965 | 0.75 | 0.71 | 0.52 | 0.48 |
| | | (0.12) | (0.11) | (0.08) | (0.08) |
| | 1973 | 0.63 | 0.86 | 0.48 | 0.50 |
| | | (0.10) | (0.14) | (0.08) | (0.08) |
| $\eta_{33}$ | 1965 | −1.70 | −1.60 | −1.48 | −1.45 |
| | | (0.19) | (0.18) | (0.13) | (0.12) |
| | 1973 | −1.59 | −1.81 | −1.45 | −1.47 |
| | | (0.16) | (0.21) | (0.12) | (0.12) |
| $\eta_{34}$ | 1965 | −0.84 | −0.79 | −0.58 | −0.54 |
| | | (0.11) | (0.11) | (0.08) | (0.07) |
| | 1973 | −0.71 | −0.96 | −0.54 | −0.56 |
| | | (0.09) | (0.13) | (0.07) | (0.07) |
| $\eta_{41}$ | 1965 | −0.50 | −0.27 | −0.42 | −0.24 |
| | | (0.10) | (0.05) | (0.08) | (0.05) |
| | 1973 | −0.46 | −0.25 | −0.38 | −0.25 |
| | | (0.09) | (0.05) | (0.07) | (0.05) |
| $\eta_{42}$ | 1965 | −0.40 | −0.21 | −0.33 | −0.19 |
| | | (0.06) | (0.03) | (0.05) | (0.03) |
| | 1973 | −0.36 | −0.19 | −0.30 | −0.20 |
| | | (0.06) | (0.03) | (0.05) | (0.03) |
| $\eta_{43}$ | 1965 | −0.60 | −0.34 | −0.51 | −0.30 |
| | | (0.08) | (0.04) | (0.07) | (0.04) |
| | 1973 | −0.56 | −0.31 | −0.47 | −0.31 |
| | | (0.07) | (0.04) | (0.06) | (0.04) |
| $\eta_{44}$ | 1965 | 0.52 | −0.17 | 0.25 | −0.25 |
| | | (0.13) | (0.07) | (0.10) | (0.06) |
| | 1973 | 0.38 | −0.25 | 0.15 | −0.24 |
| | | (0.11) | (0.06) | (0.10) | (0.06) |

| Netherlands | Norway | U.K. | U.S. | West Germany |
|---|---|---|---|---|
| 0.76 | [b] | 0.68 | 0.51 | 1.06 |
| (0.12) | | (0.11) | (0.08) | (0.17) |
| 0.21 | [b] | 0.41 | 0.71 | 0.62 |
| (0.03) | | (0.07) | (0.11) | (0.10) |
| −1.71 | [b] | −1.64 | −1.48 | −1.99 |
| (0.19) | | (0.17) | (0.12) | (0.26) |
| −1.20 | [b] | −1.39 | −1.66 | −1.58 |
| (0.05) | | (0.10) | (0.18) | (0.15) |
| −0.85 | [b] | −0.76 | −0.57 | −1.19 |
| (0.11) | | (0.10) | (0.08) | (0.16) |
| −0.24 | [b] | −0.46 | −0.79 | −0.70 |
| (0.03) | | (0.06) | (0.11) | (0.09) |
| −0.44 | −0.15 | −0.28 | −0.28 | −0.20 |
| (0.08) | (0.03) | (0.05) | (0.05) | (0.04) |
| −0.62 | −0.15 | −0.27 | −0.26 | −0.26 |
| (0.12) | (0.03) | (0.05) | (0.05) | (0.05) |
| −0.34 | −0.12 | −0.22 | −0.22 | −0.16 |
| (0.05) | (0.02) | (0.03) | (0.03) | (0.02) |
| −0.48 | −0.12 | −0.21 | −0.21 | −0.21 |
| (0.08) | (0.02) | (0.03) | (0.03) | (0.03) |
| −0.54 | −0.19 | −0.34 | −0.35 | −0.25 |
| (0.07) | (0.02) | (0.05) | (0.05) | (0.03) |
| −0.76 | −0.19 | −0.33 | −0.32 | −0.32 |
| (0.10) | (0.02) | (0.04) | (0.04) | (0.04) |
| 0.32 | −0.54 | −0.15 | −0.14 | −0.38 |
| (0.11) | (0.04) | (0.07) | (0.07) | (0.05) |
| 0.86 | −0.54 | −0.19 | −0.20 | −0.21 |
| (0.16) | (0.04) | (0.07) | (0.07) | (0.07) |

**Table 4.9**
Partial price elasticities for stationary model, all nine countries pooled together

| Elasticity[a] | Year | Belgium | Canada | France | Italy |
|---|---|---|---|---|---|
| $\eta_{11}$ | 1965 | −1.55 | [b] | −1.73 | −3.18 |
| | | (0.10) | | (0.13) | (0.38) |
| | 1973 | −2.46 | [b] | −4.40 | −12.49 |
| | | (0.25) | | (0.59) | (1.99) |
| $\eta_{12}$ | 1965 | 0.02 | [b] | 0.02 | 0.07 |
| | | (0.06) | | (0.08) | (0.24) |
| | 1973 | 0.05 | [b] | 0.12 | 0.39 |
| | | (0.16) | | (0.38) | (1.27) |
| $\eta_{13}$ | 1965 | 0.32 | [b] | 0.42 | 1.26 |
| | | (0.06) | | (0.08) | (0.24) |
| | 1973 | 0.84 | [b] | 1.96 | 6.64 |
| | | (0.16) | | (0.37) | (1.25) |
| $\eta_{14}$ | 1965 | 0.21 | [b] | 0.28 | 0.84 |
| | | (0.06) | | (0.08) | (0.23) |
| | 1973 | 0.57 | [b] | 1.32 | 4.45 |
| | | (0.16) | | (0.37) | (1.24) |
| $\eta_{21}$ | 1965 | 0.07 | 0.02 | 0.08 | 0.07 |
| | | (0.23) | (0.08) | (0.26) | (0.21) |
| | 1973 | 0.02 | 0.02 | 0.02 | 0.03 |
| | | (0.08) | (0.08) | (0.08) | (0.11) |
| $\eta_{22}$ | 1965 | −1.00 | −1.00 | −1.01 | −1.01 |
| | | (0.22) | (0.07) | (0.24) | (0.20) |
| | 1973 | −1.00 | −1.00 | −1.00 | −1.00 |
| | | (0.07) | (0.07) | (0.08) | (0.11) |
| $\eta_{23}$ | 1965 | 0.93 | 0.30 | 1.01 | 0.84 |
| | | (0.18) | (0.06) | (0.20) | (0.16) |
| | 1973 | 0.32 | 0.30 | 0.32 | 0.45 |
| | | (0.06) | (0.06) | (0.06) | (0.09) |
| $\eta_{24}$ | 1965 | −0.99 | −0.32 | −1.08 | −0.90 |
| | | (0.16) | (0.05) | (0.18) | (0.15) |
| | 1973 | −0.34 | −0.32 | −0.35 | −0.48 |
| | | (0.06) | (0.05) | (0.06) | (0.08) |
| $\eta_{31}$ | 1965 | 0.89 | 0.84 | 0.61 | 0.56 |
| | | (0.17) | (0.16) | (0.11) | (0.11) |
| | 1973 | 0.74 | 1.01 | 0.57 | 0.59 |
| | | (0.14) | (0.19) | (0.11) | (0.11) |

[a] 1 = solid, 2 = liquid, 3 = gas, 4 = electricity.
[b] Almost no solid fuel is consumed in the residential sectors of Canada and the U.S., and almost no natural gas is consumed in Norway, so that these elasticities are meaningless.

| Netherlands | Norway | U.K. | U.S. | West Germany |
|---|---|---|---|---|
| −1.93 | −3.67 | −1.79 | b | −3.84 |
| (0.16) | (0.46) | (0.14) | | (0.49) |
| −14.88 | −10.45 | −2.36 | b | −13.5 |
| (2.42) | (1.64) | (0.24) | | (2.17) |
| 0.03 | 0.09 | 0.03 | b | 0.10 |
| (0.10) | (0.29) | (0.09) | | (0.31) |
| 0.47 | 0.32 | 0.05 | b | 0.43 |
| (1.54) | (1.05) | (0.15) | | (1.38) |
| 0.54 | 1.54 | 0.46 | b | 1.64 |
| (0.10) | (0.29) | (0.09) | | (0.31) |
| 8.0 | 5.46 | 0.78 | b | 7.22 |
| (1.5) | (1.02) | (0.15) | | (1.36) |
| 0.36 | 1.03 | 0.31 | b | 1.1 |
| (0.10) | (0.29) | (0.08) | | (0.31) |
| 5.38 | 3.66 | 0.53 | b | 4.84 |
| (1.50) | (1.02) | (0.15) | | (1.35) |
| 0.03 | 0.07 | 0.14 | 0.03 | 0.04 |
| (0.10) | (0.23) | (0.46) | (0.09) | (0.13) |
| 0.04 | 0.04 | 0.16 | 0.02 | 0.03 |
| (0.15) | (0.14) | (0.52) | (0.08) | (0.09) |
| −1.00 | −1.01 | −1.01 | −1.00 | −1.00 |
| (0.10) | (0.22) | (0.44) | (0.08) | (0.13) |
| −1.01 | −1.01 | −1.02 | −1.00 | −1.00 |
| (0.14) | (0.14) | (0.49) | (0.07) | (0.08) |
| 0.40 | 0.91 | 1.82 | 0.34 | 0.53 |
| (0.08) | (0.18) | (0.35) | (0.07) | (0.10) |
| 0.58 | 0.57 | 2.04 | 0.30 | 0.35 |
| (0.11) | (0.11) | (0.39) | (0.06) | (0.07) |
| −0.43 | −0.97 | −1.94 | −0.36 | −0.57 |
| (0.07) | (0.16) | (0.32) | (0.06) | (0.09) |
| −0.62 | −0.61 | −2.17 | −0.32 | −0.37 |
| (0.10) | (0.10) | (0.36) | (0.05) | (0.06) |
| 0.89 | b | 0.80 | 0.60 | 1.25 |
| (0.17) | | (0.15) | (0.11) | (0.23) |
| 0.25 | b | 0.48 | 0.83 | 0.73 |
| (0.05) | | (0.09) | (0.16) | (0.14) |

**Table 4.9**
(continued)

| Elasticity[a] | Year | Belgium | Canada | France | Italy |
|---|---|---|---|---|---|
| $\eta_{32}$ | 1965 | 0.67 | 0.63 | 0.46 | 0.43 |
| | | (0.13) | (0.12) | (0.09) | (0.08) |
| | 1973 | 0.56 | 0.76 | 0.43 | 0.44 |
| | | (0.11) | (0.15) | (0.08) | (0.09) |
| $\eta_{33}$ | 1965 | −1.81 | −1.76 | −1.56 | −1.51 |
| | | (0.20) | (0.19) | (0.14) | (0.13) |
| | 1973 | −1.68 | −1.92 | −1.52 | −1.54 |
| | | (0.17) | (0.23) | (0.13) | (0.13) |
| $\eta_{34}$ | 1965 | −0.75 | −0.70 | −0.51 | −0.47 |
| | | (0.16) | (0.15) | (0.11) | (0.10) |
| | 1973 | −0.62 | −0.85 | −0.48 | −0.49 |
| | | (0.13) | (0.18) | (0.10) | (0.10) |
| $\eta_{41}$ | 1965 | 0.44 | 0.24 | 0.36 | 0.21 |
| | | (0.12) | (0.06) | (0.10) | (0.06) |
| | 1973 | 0.39 | 0.22 | 0.33 | 0.22 |
| | | (0.11) | (0.06) | (0.09) | (0.06) |
| $\eta_{42}$ | 1965 | −0.52 | −0.28 | −0.43 | −0.25 |
| | | (0.09) | (0.05) | (0.07) | (0.04) |
| | 1973 | −0.48 | −0.26 | −0.39 | −0.26 |
| | | (0.08) | (0.04) | (0.07) | (0.04) |
| $\eta_{43}$ | 1965 | −0.55 | −0.30 | −0.45 | −0.27 |
| | | (0.12) | (0.06) | (0.09) | (0.06) |
| | 1973 | −0.50 | −0.27 | −0.42 | −0.27 |
| | | (0.11) | (0.06) | (0.09) | (0.06) |
| $\eta_{44}$ | 1965 | −0.36 | −0.65 | −0.47 | −0.69 |
| | | (0.14) | (0.07) | (0.11) | (0.07) |
| | 1973 | −0.42 | −0.68 | −0.52 | −0.68 |
| | | (0.12) | (0.07) | (0.10) | (0.07) |

| Netherlands | Norway | U.K. | U.S. | West Germany |
|---|---|---|---|---|
| 0.67 | [b] | 0.60 | 0.45 | 0.94 |
| (0.13) | | (0.12) | (0.09) | (0.18) |
| 0.19 | [b] | 0.36 | 0.63 | 0.55 |
| (0.04) | | (0.07) | (0.12) | (0.11) |
| −1.81 | [b] | −1.73 | −1.54 | −2.14 |
| (0.20) | | (0.18) | (0.14) | (0.29) |
| −1.23 | [b] | −1.44 | −1.76 | −1.67 |
| (0.06) | | (0.11) | (0.19) | (0.17) |
| −0.75 | [b] | −0.67 | −0.50 | −1.05 |
| (0.16) | | (0.14) | (0.11) | (0.22) |
| −0.21 | [b] | −0.41 | −0.70 | −0.62 |
| (0.04) | | (0.09) | (0.15) | (0.13) |
| 0.38 | 0.13 | 0.24 | 0.25 | 0.18 |
| (0.10) | (0.04) | (0.07) | (0.07) | (0.05) |
| 0.53 | 0.13 | 0.23 | 0.23 | 0.23 |
| (0.15) | (0.04) | (0.06) | (0.06) | (0.06) |
| −0.46 | −0.16 | −0.29 | −0.29 | −0.21 |
| (0.08) | (0.03) | (0.05) | (0.05) | (0.03) |
| −0.64 | −0.16 | −0.28 | −0.27 | −0.27 |
| (0.11) | (0.03) | (0.05) | (0.05) | (0.04) |
| −0.48 | −0.16 | −0.30 | −0.31 | −0.22 |
| (0.10) | (0.03) | (0.06) | (0.07) | (0.05) |
| −0.67 | −0.17 | −0.29 | −0.29 | −0.29 |
| (0.14) | (0.03) | (0.06) | (0.06) | (0.06) |
| −0.45 | −0.81 | −0.65 | −0.64 | −0.74 |
| (0.12) | (0.04) | (0.08) | (0.08) | (0.05) |
| −0.22 | −0.81 | −0.67 | −0.66 | −0.67 |
| (0.17) | (0.04) | (0.07) | (0.07) | (0.07) |

**Table 4.10**
Total fuel price elasticities for the nonstationary fuel-choice model, all countries pooled

| Elasticity[a] | Year | Belgium | Canada | France | Italy |
|---|---|---|---|---|---|
| $\eta_{11}$ | 1965 | −1.07 | [b] | −1.12 | −1.12 |
| | 1973 | −1.06 | [b] | −1.09 | −1.08 |
| $\eta_{12}$ | 1965 | −0.05 | [b] | −0.07 | −0.18 |
| | 1973 | −0.13 | [b] | −0.30 | −0.92 |
| $\eta_{13}$ | 1965 | 0.27 | [b] | 0.34 | 1.09 |
| | 1973 | 0.74 | [b] | 1.73 | 5.94 |
| $\eta_{14}$ | 1965 | −0.26 | [b] | −0.36 | −1.03 |
| | 1973 | −0.67 | [b] | −1.55 | −5.19 |
| $\eta_{21}$ | 1965 | −0.20 | −0.06 | −0.22 | −0.16 |
| | 1973 | −0.06 | −0.05 | −0.06 | −0.08 |
| $\eta_{22}$ | 1965 | −1.21 | −1.17 | −1.27 | −1.25 |
| | 1973 | −1.11 | −1.19 | −1.15 | −1.19 |
| $\eta_{23}$ | 1965 | 1.02 | 0.31 | 1.10 | 0.91 |
| | 1973 | 0.34 | 0.31 | 0.33 | 0.47 |
| $\eta_{24}$ | 1965 | −0.76 | −0.30 | −0.85 | −0.74 |
| | 1973 | −0.27 | −0.31 | −0.29 | −0.41 |
| $\eta_{31}$ | 1965 | 0.75 | 0.75 | 0.50 | 0.49 |
| | 1973 | 0.65 | 0.91 | 0.50 | 0.52 |
| $\eta_{32}$ | 1965 | 0.74 | 0.66 | 0.50 | 0.46 |
| | 1973 | 0.60 | 0.80 | 0.44 | 0.46 |
| $\eta_{33}$ | 1965 | −1.78 | −1.79 | −1.61 | −1.58 |
| | 1973 | −1.66 | −1.95 | −1.56 | −1.59 |
| $\eta_{34}$ | 1965 | −0.86 | −0.85 | −0.61 | −0.60 |
| | 1973 | −0.72 | −1.03 | −0.57 | −0.61 |
| $\eta_{41}$ | 1965 | −0.54 | −0.27 | −0.46 | −0.26 |
| | 1973 | −0.47 | −0.25 | −0.39 | −0.25 |
| $\eta_{42}$ | 1965 | −0.40 | −0.26 | −0.34 | −0.21 |
| | 1973 | −0.38 | −0.25 | −0.34 | −0.23 |
| $\eta_{43}$ | 1965 | −0.63 | −0.35 | −0.54 | −0.33 |
| | 1973 | −0.57 | −0.33 | −0.50 | −0.34 |
| $\eta_{44}$ | 1965 | 0.43 | −0.30 | 0.12 | −0.39 |
| | 1973 | 0.31 | −0.39 | 0.04 | −0.36 |

[a] 1 = solid, 2 = liquid, 3 = gas, 4 = electricity.
[b] Almost no solid fuel is consumed in the residential sectors of Canada and the U.S., and almost no natural gas is consumed in Norway, so that these elasticities are meaningless.

| Netherlands | Norway | U.K. | U.S. | West Germany |
|---|---|---|---|---|
| −1.08 | −1.12 | −1.08 | b | −1.08 |
| −1.01 | −1.07 | −1.08 | b | −1.00 |
| −0.09 | −0.22 | −0.06 | b | −0.24 |
| −1.09 | −0.75 | −0.11 | b | −0.98 |
| 0.46 | 1.38 | 0.39 | b | 1.46 |
| 7.17 | 4.91 | 0.67 | b | 6.48 |
| −0.44 | −1.30 | −0.39 | b | −1.32 |
| −6.21 | −4.31 | −0.64 | b | −5.61 |
| −0.09 | −0.17 | −0.35 | −0.06 | −0.10 |
| −0.10 | −0.10 | −0.38 | −0.05 | −0.06 |
| −1.14 | −1.26 | −1.34 | −1.16 | −1.17 |
| −1.15 | −1.20 | −1.38 | −1.14 | −1.10 |
| 0.43 | 1.02 | 2.03 | 0.35 | 0.58 |
| 0.61 | 0.64 | 2.26 | 0.32 | 0.38 |
| −0.35 | −0.84 | −1.50 | −0.32 | −0.48 |
| −0.48 | −0.55 | −1.68 | −0.29 | −0.30 |
| 0.77 | b | 0.69 | 0.53 | 1.11 |
| 0.22 | b | 0.41 | 0.75 | 0.65 |
| 0.73 | b | 0.67 | 0.46 | 1.04 |
| 0.20 | b | 0.40 | 0.66 | 0.60 |
| −1.80 | b | −1.72 | −1.58 | −2.09 |
| −1.28 | b | −1.47 | −1.77 | −1.64 |
| −0.87 | b | −0.76 | −0.61 | −1.24 |
| −0.25 | b | −0.50 | −0.84 | −0.72 |
| −0.46 | −0.16 | −0.30 | −0.28 | −0.21 |
| −0.61 | −0.15 | −0.28 | −0.26 | −0.26 |
| −0.37 | −0.13 | −0.22 | −0.26 | −0.18 |
| −0.49 | −0.14 | −0.21 | −0.24 | −0.22 |
| −0.55 | −0.18 | −0.35 | −0.37 | −0.26 |
| −0.80 | −0.18 | −0.35 | −0.34 | −0.33 |
| 0.23 | −0.68 | −0.24 | −0.25 | −0.48 |
| 0.78 | −0.65 | −0.28 | −0.30 | −0.26 |

**Table 4.11**
Summary of elasticity estimates

| Elasticity | Estimate | Table reference |
| --- | --- | --- |
| Aggregate energy use: price elasticities | $\eta_{EE}$ = −1.05 to −1.15<br>$\eta_{EA}$ = 0.06 to 0.15<br>$\eta_{ED}$ = 0.05 to 0.13<br>$\eta_{EF}$ = 0.31 to 0.82<br>$\eta_{ET}$ = −0.17 to −0.47<br>$\eta_{ER}$ = −0.19 to −0.51 | 4.4 |
| Fuels: own-price elasticities, partial | coal: −0.94 to −0.99<br>oil: −1.04 to −1.29<br>gas: −1.20 to −1.99<br>electricity: positive to −0.54 | 4.8 |
| Fuels: own-price elasticities, total | coal: −1.00 to −1.12<br>oil: −1.10 to −1.38<br>gas: −1.28 to −2.09<br>electricity: positive to −0.68 | 4.10 |

**Table 4.12**
Alternative estimates of residential energy demand elasticities

| Elasticity | Country | Estimate | Source |
|---|---|---|---|
| Aggregate energy use: own-price elasticity | U.S. | short run: −0.12<br>long run: −0.50 | a |
| | U.S. | short run: −0.16<br>long run: −0.63 | b |
| | U.S. | −0.28 | c |
| | U.S. | −0.40 | d |
| | U.S. | short run: −0.50<br>long run: −1.70 | e |
| | Canada | −0.33 to −0.56 | f |
| | Norway | −0.30 | g |
| | West Germany | short run: −0.35<br>long run: −0.78 | e |
| | Italy | short run: −0.63<br>long run: −1.30 | e |
| | Netherlands | short run: −0.42<br>long run: −1.30 | e |
| | U.K. | short run: −0.38<br>long run: −0.42 | e |
| | 6 countries pooled | −0.71 | e |
| | 20 OECD countries pooled | −0.42 | h |
| Aggregate energy use: income elasticity | U.S. | short run: 0.10<br>long run: 0.60 | a |
| | U.S. | short run: 0.20<br>long run: 0.80 | b |
| | U.S. | 0.27 | c |
| | Canada | 0.83 to 1.26 | f |
| | Norway | 1.08 | g |

[a] Joskow and Baughman (102).
[b] Baughman and Joskow (11).
[c] Nelson (122).
[d] Jorgenson (98).
[e] Nordhaus (123).
[f] Fuss and Waverman (62).
[g] Rødseth and Strøm (145).
[h] Adams and Griffin (2).
[i] Halvorsen (76).
[j] Liew (111).
[k] Berndt and Watkins (21).
[l] Hirst, Lin, and Cope (84).
[m] Griffin (68).
[n] Mount, Chapman, and Tyrrell (121).

| Elasticity | Country | Estimate | Source |
|---|---|---|---|
| | 6 countries pooled | 1.09 | e |
| | 20 OECD countries pooled | 1.51 | h |
| Fuel consumption: partial own-price elasticities | U.S. | electricity: −0.06 (short run)<br>0.52 (long run) | m |
| | U.S. | electricity: −0.14 (short run)<br>−1.22 (long run) | n |
| | Canada | gas and oil: −0.96<br>−0.34 | f |
| | Norway | electricity: −0.22 to −0.60 | g |
| | 20 OECD countries pooled | gas: −1.05<br>oil: −0.33<br>coal: −0.81 | h |
| Fuel consumption: total on price elasticities | U.S. | electricity: −1.0 to −1.2 | i |
| | U.S. | gas: −0.15 (short run)<br>−1.01 (long run)<br>oil: −0.18 (short run)<br>−1.10 (long run)<br>electricity: −0.19 (short run)<br>−1.00 (long run) | a |
| | U.S. | gas: −1.34<br>oil: −1.89<br>electricity: −1.13 | b |
| | U.S. | gas: −1.28 to −1.77<br>electricity: −0.40 | j |
| | U.S. | gas: −0.91<br>oil: −0.91<br>electricity: −0.84 | l |
| | Canada | gas: −0.15 (short run)<br>−0.69 (long run) | k |
| | 20 OECD countries pooled | gas: −1.11<br>oil: −0.52<br>coal: −0.98 | h |

**Table 4.13**
Parameter estimates for static logit models

| Parameter | (1) Linear in relative price and income | | (2) Linear in relative price, income, and temperature | |
|---|---|---|---|---|
| $a_{14}$ | 6.7256 | (10.10) | 8.6567 | (8.75) |
| $a_{24}$ | 0.9400 | (1.55) | 0.7536 | (0.75) |
| $a_{34}$ | 1.1740 | (2.15) | −0.5394 | −0.6335 |
| $CN_{14}$ | −4.8884 | (−20.48) | −5.4918 | (−16.35) |
| $CN_{24}$ | −0.7109 | (−2.69) | −0.6725 | (−1.82) |
| $CN_{34}$ | −1.2883 | (−5.79) | −0.7377 | (−2.40) |
| $FR_{14}$ | −0.8825 | (−5.22) | −0.6833 | (−3.63) |
| $FR_{24}$ | −0.7151 | (−4.04) | −0.7428 | (−3.71) |
| $FR_{34}$ | −0.2900 | (−1.85) | −0.4983 | (−2.89) |
| $IT_{14}$ | −3.5782 | (−12.82) | −3.1205 | (−9.09) |
| $IT_{24}$ | −1.4759 | (−5.89) | −1.5400 | (−4.60) |
| $IT_{34}$ | −0.8752 | (−3.99) | −1.3700 | (−4.80) |
| $ND_{14}$ | −1.7246 | (−9.47) | −1.7623 | (−9.75) |
| $ND_{24}$ | 0.1651 | (0.17) | 0.1580 | (0.91) |
| $ND_{34}$ | 0.1263 | (0.85) | 0.1705 | (1.16) |
| $NR_{14}$ | −4.0201 | (−11.71) | −4.5917 | (−11.25) |
| $NR_{24}$ | −2.3643 | (−6.49) | −2.3383 | (−5.36) |
| $NR_{34}$ | −6.5012 | (−18.11) | −5.9818 | (−14.73) |
| $UK_{14}$ | −1.7562 | (−6.32) | −1.6575 | (−5.90) |
| $UK_{24}$ | −2.1243 | (−7.96012) | −2.1729 | (−7.90) |
| $UK_{34}$ | −0.9826 | (−4.27) | −1.1300 | (−4.81) |
| $US_{14}$ | −8.7050 | (−49.24) | −8.5911 | (−47.53) |
| $US_{24}$ | −0.7155 | (−3.73) | −0.7454 | (−3.73) |
| $US_{34}$ | −0.4900 | (−2.49) | −0.6236 | (−3.14) |
| $WG_{14}$ | −3.5738 | (−12.64) | −3.8422 | (−13.10) |
| $WG_{24}$ | −1.3589 | (−5.24) | −1.3400 | (−4.83) |
| $WG_{34}$ | −1.7872 | (−7.71) | −1.5866 | (−6.53) |
| $b_1$ | $5.4863 \times 10^{-6}$ | (0.10) | $-8.3737 \times 10^{-6}$ | (−0.15) |
| $b_2$ | $7.1110 \times 10^{-6}$ | (0.15) | $1.6490 \times 10^{-5}$ | (0.35) |
| $b_3$ | $3.4942 \times 10^{-5}$ | (2.10) | $3.3989 \times 10^{-5}$ | (2.08) |
| $b_4$ | $4.8882 \times 10^{-5}$ | (4.03) | $4.9800 \times 10^{-5}$ | (4.10) |
| $CNSD$ | | | | |
| $USSD$ | | | | |
| $NRGD$ | | | | |
| $WGGD$ | | | | |

| (3) Linear in relative price and income; dummy variables | | (4) Linear in logs of relative price and income; dummy variables | |
|---|---|---|---|
| 4.6183 | (6.34) | 26.43 | (4.55) |
| −1.1002 | (−1.66) | −13.03 | (−2.35) |
| 1.9594 | (3.25) | 8.68 | (1.71) |
| 0.3183 | (0.26) | 40.3656 | (4.08) |
| 0.1322 | (0.48) | −0.0022 | (−0.01) |
| −2.04 | (−8.22) | −1.8879 | (−6.60) |
| −0.4874 | (−2.92) | −0.3739 | (−2.35) |
| −0.3500 | (−1.95) | −0.1395 | (−0.82) |
| 0.0022 | (0.01) | 0.1220 | (0.08) |
| −2.8435 | (−9.77) | −2.4560 | (−8.16) |
| −0.7820 | (−2.95) | −0.3213 | (−1.14) |
| −0.9612 | (−4.13) | −0.6138 | (−2.54) |
| −1.366 | (−7.60) | −1.4059 | (−8.19) |
| 0.5675 | (3.27) | 0.5661 | (3.45) |
| 0.2095 | (1.41) | 0.3520 | (2.41) |
| −2.8467 | (−7.73) | −2.5151 | (−5.16) |
| −1.0996 | (−2.82) | −1.0664 | (−2.15) |
| −7.9598 | (−15.69) | −23.82 | (−5.81) |
| −0.9859 | (−3.39) | −0.56 | (−1.87) |
| −1.2533 | (−4.47) | −1.3266 | (−4.68) |
| −0.7736 | (−3.18) | −0.4693 | (−1.85) |
| −12.00 | (−9.67) | −21.3229 | (−2.03) |
| −0.3594 | (−1.91) | −0.4493 | (−2.55) |
| −1.7326 | (−7.77) | −1.9778 | (−6.23) |
| −2.8073 | (−9.51) | −2.3900 | (−7.81) |
| −0.6365 | (−2.32) | 0.1648 | (−0.56) |
| −4.3407 | (−9.73) | −24.50 | (−6.18) |
| $-4.0938 \times 10^{-6}$ | (−0.07) | 0.4894 | (1.79) |
| $-4.5742 \times 10^{-5}$ | (−1.07) | 0.4152 | (2.50) |
| −0.0002 | (−6.70) | −1.2639 | (−5.15) |
| $4.7887 \times 10^{-6}$ | (0.36) | −0.1843 | (−0.47) |
| −0.0007 | (−3.88) | −5.060 | (−4.49) |
| 0.0005 | (2.88) | 1.4289 | (1.20) |
| 0.0002 | (6.56) | 1.9984 | (4.63) |
| 0.0003 | (7.36) | 2.5392 | (5.87) |

**Table 4.13**
(continued)

| Parameter | (1) Linear in relative price and income | | (2) Linear in relative price, income, and temperature | |
|---|---|---|---|---|
| $c_{14}$ | −0.0023 | (−19.88) | −0.0023 | (−19.67) |
| $c_{24}$ | 0.0002 | (1.84) | 0.0002 | (1.67) |
| $c_{34}$ | $-4.9263 \times 10^{-5}$ | (−0.49) | $-8.2384 \times 10^{-5}$ | (0.83) |
| $d_{14}$ | | | −0.0484 | (2.08) |
| $d_{24}$ | | | 0.0059 | (0.27) |
| $d_{34}$ | | | 0.0465 | (2.65) |
| Equation 1, $R^2$ | 0.9886 | | 0.9891 | |
| Equation 2, $R^2$ | 0.7178 | | 0.7164 | |
| Equation 3, $R^2$ | 0.9543 | | 0.9568 | |

| (3) Linear in relative price and income; dummy variables | | (4) Linear in logs of relative price and income; dummy variables | |
|---|---|---|---|
| −0.0020 | (−17.84) | −4.2573 | (−15.97) |
| 0.0005 | (4.16) | 0.9737 | (3.89) |
| $-6.0993 \times 10^{-5}$ | (−0.61) | 0.1266 | (0.56) |
| 0.9909 | | 0.9901 | |
| 0.7420 | | 0.7549 | |
| 0.9592 | | 0.9574 | |

# 5 The Industrial Demand for Energy

We saw in the first two chapters of this book how the characteristics of the demand for energy in the industrial sector will depend upon the role of energy as a factor of production and its substitutability with other factors, as well as the substitutability of individual fuels within the energy aggregate. These aspects of the structure of industrial energy demand should be captured by the two-stage model developed in chapter 2. Here we present the statistical results from the estimation of the various versions of our industrial demand model and discuss the implications of the statistical evidence for the characteristics and likely future behavior of industrial energy demand.

Let us again review our procedure for applying the translog cost function to our two-stage model of industrial energy demand. We begin by estimating the fuel-share equations

$$S_i = \alpha_i + \sum_j \gamma_{ij} \log P_j, \tag{5.1}$$

where the $P_j$ are the prices of coal, oil, natural gas, and electricity. These equations are estimated subject to the parameter restrictions $\Sigma_i\alpha_i = 1$, $\gamma_{ij} = \gamma_{ji}$, and $\Sigma_i\gamma_{ij} = \Sigma_j\gamma_{ij} = 0$, but we also test the additional restrictions $\gamma_{ij} = 0$. Note that there are four fuels, but only three share equations need to be estimated since the parameters of the fourth are determined by the adding-up constraint $\Sigma S_i = 1$. In estimating these equations, we allow $\alpha_i$ to vary across countries, but we also test the restriction that they are the same for all countries (regional homogeneity). The resulting parameter estimates are used to calculate partial price elasticities (that is, elasticities based on the constancy of energy expenditures) for each of the four fuels. In addition, these parameter estimates are used to obtain an aggregate price index for energy in the industrial sector. Recall that this is done by choosing the parameter $\alpha_0$ in equation (2.44) so that the price of energy is equal to 1 in the U.S. in 1970. A relative price index is then calculated for every country in every year.

Next the factor share equations

$$S_i = \alpha_i + \gamma_{Qi} \log Q + \sum_j \gamma_{ij} \log P_j \tag{5.2}$$

are estimated, with $i$ and $j$ equal to capital, labor, and energy. Only two equations need be estimated since the third is determined from the adding-up constraint. The estimated aggregate price index for energy is used as an instrumental variable for the price of energy. The share equations are estimated in stages, with additional parameter restrictions imposed and tested at each stage. Initially only the restrictions implied by neoclassical production theory are imposed, namely, $\Sigma_i\alpha_i = 1$, $\Sigma_i\gamma_{Qi} = 0$, $\gamma_{ij} = \gamma_{ji}$, and $\Sigma_i\gamma_{ij} = \Sigma_j\gamma_{ij} = 0$. Next we test the homotheticity restrictions $\gamma_{Qi} = 0$, and then the restrictions that $\gamma_{ij} = 0$, that is, that the elasticities of substitution between all three factors are equal to 1. In addition, we test restrictions of regional homogeneity, that is, that the first-order parameters are the same across countries. After estimating these equations the resulting parameters are used to obtain elasticities of substitution and demand elasticities for the three factors, and, together with the partial fuel price elasticities obtained earlier, to obtain total elasticities of demand for each of the four fuels.

Ten countries are included in our sample: Canada, France, Italy, Japan, the Netherlands, Norway, Sweden, the U.K., the U.S., and West Germany. The data for all of these countries span the period 1959 to 1974.

In estimating the fuel share equations (5.1) and the factor share equations (5.2) a number of choices must be made regarding the pooling of data and the choice of time bounds. First, regional dummy variables can be used to allow the intercept parameters $\alpha_i$ of the share equations to vary across countries, or to allow the second-order parameters $\gamma_{ij}$ to vary across countries. We found the latter alternative to involve a considerable reduction in degrees of freedom, so that the resulting elasticities had large standard errors. Therefore we report here only the use of regional dummy variables for the intercept terms. We also consider using no dummy variables at all (so that the energy cost function is regionally homogeneous), and we use the standard chi-square test to determine whether this restriction is warranted.

Second, fuel prices in Canada and the U.S. have for many years been significantly lower than in the other eight countries in our sample, and this might have resulted in a production structure different enough to suggest pooling these countries separately. We therefore estimate models in which all ten countries are pooled together, and models in which Canada and the U.S. are pooled separately.

Third, although our data span the period 1959 to 1974, there is a question as to whether the 1974 data should be considered to have come from the same population as the 1959 to 1973 data. Energy prices rose rapidly in 1974, but the use of energy can respond only slowly to such a change. It is therefore not clear whether the 1974 data point lies on the same long-run cost function as the earlier data. We estimate models both including and excluding the 1974 data, and by examining the sensitivity of the estimates to the 1974 data, determine whether prices and shares moved off the long-run cost function in 1974.

Finally, it is useful to test whether we are indeed estimating long-run cost functions. To do this we estimate our models using data at three-year intervals, and compare the results to the corresponding ones obtained using annual data. If the resulting estimates are nearly the same, we can conclude that we have indeed estimated a long-run cost function.

## 5.1 Estimation of the Translog Fuel-Share Model

Of the various versions of this model that have been estimated, eight are reported here in table 5.1. (Standard errors are in parentheses.) In the first version all ten countries are pooled and the 1959 to 1973 data are used, but the model is restricted to be regionally homogeneous, that is, no intercept dummy variables are included. We test the restriction of regional homogeneity by comparing the model to the equivalent unrestricted model of column 6. The value of the chi-square statistic is 628, and given that there are 27 parameter restrictions, this is well above the critical 1 percent level of 45. We therefore include intercept dummy variables in all other versions of the model. (A model is also estimated using regional dummy variables for the second-order terms, but we do not report the estimated parameters here. The resulting own-price elasticities, however, are shown in table 5.4, and as can be seen from that table, many of the elasticities are statistically insignificant.)

In columns 2, 3, 4, and 5 Canada and the U.S. are pooled separately, and in columns 6 and 7 they are pooled together with the European countries and Japan. Also, the results of including or excluding the 1974 data can be seen by comparing columns 2 and 3, columns 4 and 5, and columns 6 and 7. Note that in all cases the $\beta_{ij}$ estimates change considerably when the 1974 data point is added, leading us to believe that it should not be included with the 1959 to 1973 sample. Also, by comparing columns 2, 4, and 6, we see that the $\beta_{ij}$ parameter estimates for Canada and the U.S. are quite different from those for Japan and the European countries when the former are pooled separately. In addition, the parameter estimates for Canada and the U.S. are still statistically significant when the pooling is separate. We take this as an indication that it is preferable to pool Canada and the U.S. separately.

By comparing column 8 with column 4, we see that the use of data at three-year intervals results in little change in the estimated parameters, although the standard errors of the estimates become larger. The resulting price elasticities also do not change much (compare tables 5.2 and 5.5), so that we conclude that we are indeed estimating long-run elasticities.

Estimated elasticities are shown in tables 5.2, 5.3, 5.4, and 5.5, with standard errors in parentheses. Recall that these standard errors are asymptotic approximations of the true standard errors, and are calculated under the assumption that the expenditure shares are constant and equal to the means of their estimated values. Formulas for these standard errors were given in equation (2.57).

Table 5.2 shows partial own- and cross-price elasticities for Canada and the U.S. pooled separately, for both the 1959 to 1973 and 1959 to 1974 time bounds. The same elasticities are shown in table 5.3 for all ten countries pooled together. Note that while elasticities for solid fuel and electricity are more or less the same for the four different versions of the model, elasticities for liquid fuel (largely residuel fuel oil) and natural gas vary across the four versions. In table 5.3 we see that pooling all ten countries results in own-price elasticities for liquid fuel ($\eta_{22}$) and natural gas ($\eta_{33}$) that vary little across countries. We see in table 5.2 that pooling Canada and the U.S. separately results in the oil elasticity becoming much larger for these two countries and smaller for the other countries, while the opposite is true for the natural gas elasticity. In addition, the own- and cross-price elasticities for oil are statistically significant for Canada and the U.S. in table 5.2, whereas many of them are insignificant in table 5.3. This is a further indication that Canada and the U.S. should be pooled separately.

Note in table 5.2 that including the 1974 data results in a large change in the price elasticities for oil, and in particular these elasticities become smaller and in many cases insignificant. This is not surprising. Oil prices rose considerably in that year, but demand can adjust to these higher prices only slowly. We should therefore expect that the 1974 data lies on a short-run cost function, and including this data would give us demand elasticities for oil that are somewhere between the short and long run. This, however, is of little value given that we are estimating a static cost function. By leaving out the 1974 data, we can more safely assume that our estimates represent long-run elasticities. We therefore choose as our preferred model version (a) in table 5.2—that is, Canada and the U.S. pooled separately, with the estimation done over the 1959 to 1973 time bounds.

Note that except for electricity, the elasticities of our preferred model are large in magnitude. (Remember that these are partial price elasticities, in other words, based on constant energy consumption, so that the total own-price elasticities will be even larger in magnitude.) Own-price elasticities for coal range from −1 in France and West Germany to about −2 in Norway, Canada, and the U.S. (where coal has had a smaller share of industrial energy consumption). Own-price elasticities for natural gas are less than −1 in the European countries and Japan, but −0.33 in Canada and −0.52 in the U.S. This is reasonable given that natural gas prices have been much lower in Canada and the U.S. than in the rest of the world. On the other hand, Canada and the U.S. have the largest own-price elasticities for oil, even though they had relatively low prices. We can only explain this on the basis of a greater availability of alternative fuels at low prices (notably natural gas), so that producers chose technologies allowing for greater interfuel substitution possibility. Finally, note that the own-price elasticities for electricity are all quite small in magnitude. This is not surprising: since electricity is a much more expensive fuel on a thermal equivalent basis, we expect it to be used only where there is little or no possibility of using an alternative fuel.

## 5.2 The Aggregate Price Index for Energy

We generate our aggregate price index for energy using the estimated parameters from the preferred fuel-share model, that is, the version in which the U.S. and Canada are pooled separately, and the data span the period 1959 to 1973. To compute the price index, we apply the estimated parameters from columns 2 and 4 of table 5.1 to the energy cost function:

$$\log P_E = \alpha_0 + \sum_i \alpha_i \log P_i + \sum_i \sum_j \gamma_{ij} \log P_i \log P_j. \tag{5.3}$$

Note that the first-order parameters $\alpha_i$ will vary across countries. We choose $\alpha_0$, the unobservable parameter of the equation, so that the price of energy $P_E$ is equal to 1.0 in the U.S. in 1970. Then using the data on fuel prices, we can generate the relative energy price index over time for each country. The resulting energy price indices are shown in table 5.6. These indices will in turn serve as the instrumental variable for the price of energy in the estimation of the factor-share model.

Since the energy price index is an important piece of data in the estimation of the factor-share model, it is useful to examine it in more detail and to compare its behavior across countries. In figure 5.1 the price index is plotted over time for each of five countries, France, the Netherlands, the U.K., the U.S., and West Germany. Observe from this figure and from table 5.6 that the variation across countries is considerable; even if the Netherlands is omitted, the index in some countries is three or four times as large as in others. On the other hand, for most countries there is very little variation

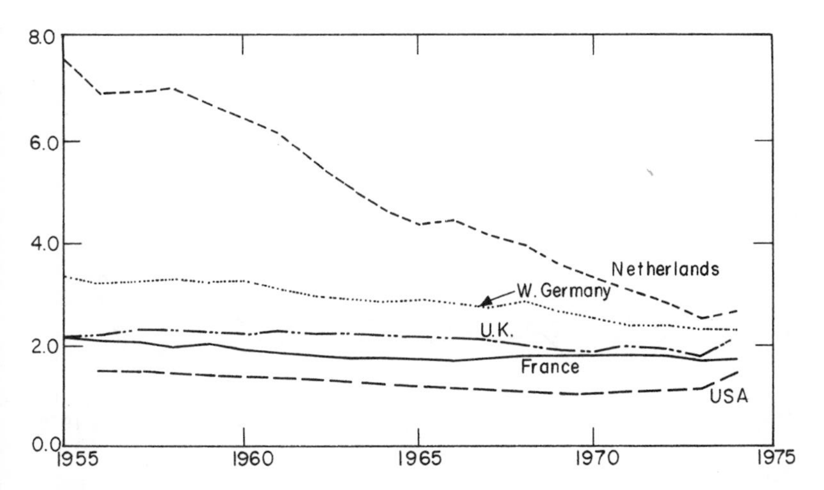


**Figure 5.1**
Energy price index (U.S. = 1 in 1970)

in the index across time, at least until 1973. (The annual percent change in the index for three countries, France, the U.S., and West Germany, is shown in figure 5.2.) This means that most of the explanation provided by this data will come from intercountry differences, which reinforces our expectation that we are estimating a long-run cost function.

The patterns of industrial energy use across countries are similar to those of energy prices. In figure 5.3 total energy requirements per 1970 dollar of manufacturing output is plotted over time for the same five countries as in figure 5.1. Observe that there is considerable intercountry variation in energy use per dollar of output (although the variation is not as great in percentage terms as it is for the price of energy). These intercountry differences in prices and quantities should enable us to obtain reasonably low-variance estimates of the long-run demand elasticities for energy and other factors of production.

### 5.3 Estimation of the Factor-Share Model

We turn now to the model of expenditure shares for capital, labor, and energy. Once again, choices must be made regarding the use

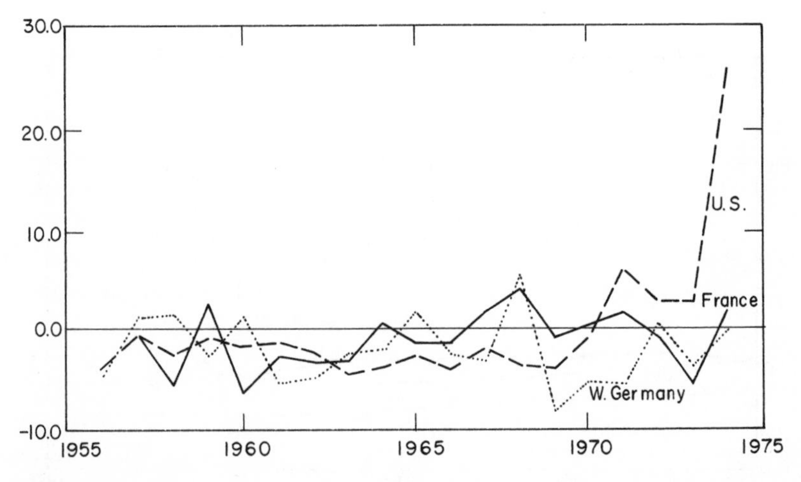


**Figure 5.2**
Annual percent change in energy price index

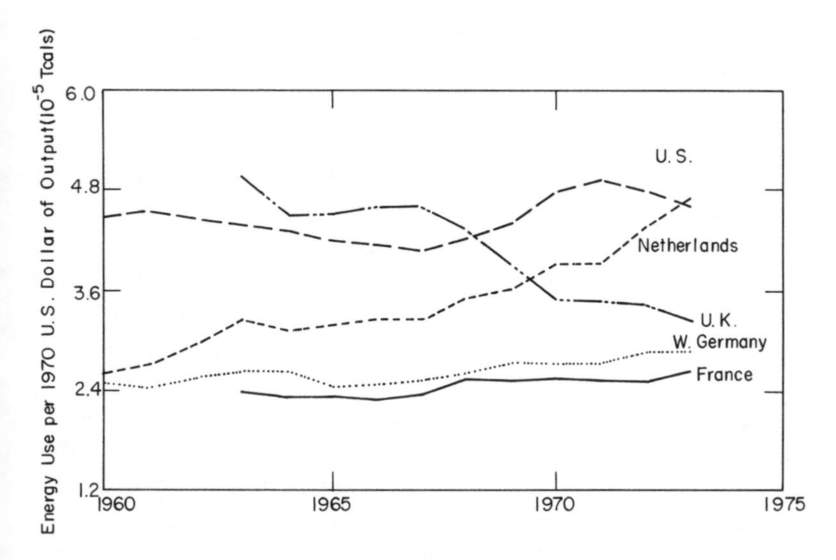


**Figure 5.3**
Total energy requirements per dollar of manufacturing output

of regional dummy variables and the pooling of data.[1] In addition, we cannot a priori assume homotheticity but must test this as a possible restriction on the model.

Parameter estimates for several forms of the model are given in table 5.7, although other forms were estimated as well. In columns 1 and 2 the share equations have been estimated without the inclusion of regional dummy variables. In column 1 the cost function is homothetic ($\gamma_{Qi} = 0$, $i = K, L, E$), and in column 2 it is nonhomothetic. These two models can be used as a first test for homotheticity; the value of the test statistic is 68.4, and this is significant at the 1 percent level, indicating that homotheticity cannot be accepted. In column 3 the cost function is nonhomothetic, and intercept dummy variables are added to the share equations. A comparison of this model with that in column 2 allows us to test for constancy of the first-order parameters across countries. The test statistic is 359.8, which is significant at the 1 percent level (27

1. Inclusion of 1974 data is not a question here; our derived data for the price index of capital services extend only through 1973.

degrees of freedom), so that intercept dummy variables are retained in the share equations.

In columns 4 and 5 a nonhomothetic cost function is again estimated with intercept dummy variables in the share equations, but now Canada and the U.S. are pooled separately. Note that the $\gamma_{ij}$ parameter values change considerably, with the values in column 3 generally midway between those in columns 4 and 5. In addition, the parameter estimates generally become more significant for Europe and Japan but less significant for Canada and the U.S. This provides us with no firm indication of whether Canada and the U.S. should be pooled separately.

A second test for homotheticity is performed by estimating homothetic versions of the cost functions in columns 4 and 5 (the resulting parameter estimates are not reported here). These models can then be compared to the corresponding nonhomothetic models. For Europe and Japan the test statistic is 8.58, and this is significant at the 2.5 percent level, so that homotheticity cannot be accepted here. For Canada and the U.S. the test statistic is 1.25, which is not significant at the 10 percent level, so that homotheticity could be accepted. However, if the regional dummy variables are eliminated, the same test for homotheticity gives a test statistic of 32.4, which is significant at the 1 percent level. In addition, testing constancy of the first-order terms in the nonhomothetic model gives a test statistic of 2.35 which is not significant at the 10 percent level. We therefore consider the test for homotheticity to be inconclusive, and retain a nonhomothetic model.[2]

In column 6 the nonhomothetic cost function is estimated, with all ten countries pooled, using data at three-year intervals. Comparing columns 3 and 6 we see that many of the estimated parameter values do change considerably. We are thus less confident that we have estimated a long-run cost function than we are for the fuel-share model.

Elasticities of substitution for the three factors are shown in table 5.8 for the model in which all ten countries are pooled together (column 3), and the model in which Canada and the U.S. are pooled

2. We also tested the hypotheses that the $\gamma_{ij}$ are all zero, that is, that the production function is Cobb-Douglas. The test statistics are significant at the 1 percent level, so that these hypotheses can be rejected.

separately (columns 4 and 5). Price elasticities of demand for the three factors are shown in table 5.9, again for the model in which all ten countries are pooled and for the model in which Canada and the U.S. are pooled separately. Note that the choice of pooling method has little effect on the estimated elasticities for the European countries and Japan, but almost all of the elasticities for Canada and the U.S. are larger in magnitude when these countries are pooled separately. Pooling Canada and the U.S. separately, however, also results in elasticities with much larger standard errors. We are therefore inclined to choose as a preferred model that for which all of the countries are pooled together.

Let us now consider the implications of these elasticity estimates for energy demand and the substitutability of energy with other factors of production. First we see that the elasticity of substitution for energy and capital ($\sigma_{KE}$) is positive, although small (0.61 to 0.86). We thus find that energy and capital are substitutes, and not complements as earlier studies for the U.S. had indicated.[3] We find that labor and energy are also substitutes (with elasticities of substitution close to 1); this is not surprising, and is supported by most other work. Similarly, we find that capital and labor are substitutes, as expected.

The own-price elasticities of demand for capital and labor are around $-0.3$ to $-0.5$, which is in agreement with most earlier work. The own-price elasticities for energy, however, are larger in magnitude than most of those obtained earlier by others. We find this elasticity to be about $-0.8$, whereas most earlier estimates were around $-0.4$ to $-0.5$. Much of the earlier work, however, used data for a single country rather than pooled international data, and therefore was more likely to have captured short- or intermediate-run elasticity.[4] Finally, note from the cross-price elasticities of energy and capital that over the long run a doubling in the price of energy should result in a 5 percent increase in the demand for capital and

3. Berndt and Wood (22), for example, found complementarity between energy and capital. Griffin and Gregory (71), using pooled international data at four-year intervals, obtain results very similar to ours. Later in this chapter we offer alternative interpretations of the different findings.

4. Our estimate is close to that found by Griffin and Gregory (71), who also used international data.

a 6 percent or 7 percent increase in the demand for labor as substitution away from energy takes place.

Since our model is nonhomothetic, the elasticity of energy demand with respect to output changes ($\eta_{EQ}$) need not be equal to 1. We can calculate this elasticity, and also determine what our estimated cost function implies about economies of scale in aggregate production. In table 5.10 we show the index of scale economies (SCE), introduced by Christensen and Greene (38), and the elasticity of energy demand with respect to output changes, as given by equation (2.53). (The indices and elasticities for each country are calculated at the point of means.) Note that the index of scale economies is insignificantly different from zero for each country, so that the aggregate cost functions exhibit nearly constant returns to scale. The output elasticity of energy demand, however, is significantly less than unity. Thus, even if energy prices remain constant relative to other prices, as output increases there will be substitution away from energy.[5]

We can now examine the total price elasticities for the individual fuel demands. The derivations of these elasticities are given by equations (2.48) and (2.49). The calculated values of these elasticities are shown in table 5.11, and they are based on own-price elasticities of energy demand obtained from table 5.9 with all ten countries pooled together, and partial fuel price elasticities obtained from table 5.2. Note that these total elasticities are larger than the partial elasticities of tables 5.2 and 5.3, since they account for changes in the use of energy as well as interfuel substitution.

We find that coal has the largest own-price elasticities, ranging from $-1.29$ to $-2.24$. For Europe and Japan, own-price elasticities for natural gas are large ($-1.37$ to $-2.34$), while those for oil are small ($-0.6$ to $-.56$). We attribute this to the fact that for two countries (Netherlands and West Germany), as oil and gas prices fell, there was a large increase in the share of natural gas (from almost zero) as supplies became available for the first time. This might have tended to bias the natural gas elasticities upwards. It is more difficult to explain the low oil price elasticities; oil prices on

5. This result is not surprising given the data. Note from table 3.6 that for large-output countries like the U.S., the share of energy is smaller than for low-output countries.

a thermal-equivalent basis were generally the lowest of any fuel, but oil did not gain a dominant share in Europe. For Canada and the U.S. (which were pooled separately in the fuel-choice model) the situation is reversed—the price elasticities for oil are larger than those for natural gas. Here natural gas prices were lower than oil prices, and the share of oil was small (in the U.S.) and roughly constant over time.

In table 5.12 we show the elasticities of the average cost of output with respect to the price of energy and the prices of the individual fuels. These elasticities are given by equations (2.55) and (2.56). They are presented here for two versions of the factor share model—Canada and the U.S. pooled separately, and all ten countries pooled together. These elasticities give us the effects of increases in energy prices on the cost of output, assuming the level of output stays fixed. They thus provide information about the inflationary impact of an energy price rise, but they do not provide information about the effect on the level of GNP. Note that in the U.S. a 10 percent increase in the cost of energy would result in about a 0.3 percent increase in the cost of output, whereas for Italy, Japan, and Sweden the cost of output would rise about 0.7 percent. However, in comparing these elasticities across countries, remember that they are dependent on the shares of energy, and the fuel shares. Thus Italy, Japan, and Sweden have the largest values of $\eta_{CE}$ because they have the largest shares of energy in the cost of output.

## 5.4 Summary of Results and Comparison with Other Studies

We have seen that the use of a two-stage weakly separable cost function can indeed provide a means of estimating demand elasticities for aggregate industrial energy use and for individual fuels. In addition, by pooling international time-series cross-section data we were able to obtain a sample large enough to provide low-variance estimates of essentially long-run elasticities. A summary of some of the more important elasticity estimates that were discussed in this chapter is given in table 5.13.

Perhaps our most important result is that the own-price elasticity of aggregate industrial energy demand appears to be significantly larger than had been thought previously and energy and capital appear to be substitutable rather than complementary factors of production. We attribute this result to the long-run nature of our estimates.

We found the total own-price elasticities of coal and natural gas to be large, as expected, so that there is considerable room for interfuel substitution here. On the other hand, we found the total own-price elasticities of oil and electricity to be below 1 in magnitude, at least for Europe and Japan. While we would expect to observe the small elasticity for electricity (there is little flexibility in its use), it is harder to justify the elasticities for oil.

We also found that the aggregate cost functions are mildly, but significantly, nonhomothetic, so that the elasticity of aggregate energy use with respect to output changes is below 1 (generally around 0.7 or 0.8). This is not due to economies of scale in the long run (we found the cost functions to exhibit approximately constant returns to scale) but rather to substitution away from energy as output increases. Finally, we found that further increases in the price of energy would have only a small (long-run) impact on the total cost of production. This is due in part to energy's small share in production, and in part to substitution possibilities.

Let us now compare our results to those obtained by others. In table 5.14 we present a survey of recent estimates of industrial energy demand elasticities. We can compare these estimates with those that we have obtained, determine what kind of consensus estimates could be arrived at for various elasticities, and consider how and why our estimates might differ from these.

Note that there is mixed evidence on the substitutability of energy and capital. Berndt and Wood (22), Fuss (60), and Magnus (114) find energy and capital to be strong complements, but they worked with time-series data for a single country, and might have estimated a short-run cost or production function. Halvorsen and Ford (79) and Fuss and Waverman (62) obtain mixed results on energy-capital substitutability, depending on the particular disaggregated industry or the particular form of the cost function. Only Griffin and Gregory (71) find strong evidence of capital-energy substitutability, and their estimate of the Allen elasticity of substitution is close to ours (1.01 compared to about 0.8). This is reassuring since both their study and ours use international data and presume to estimate long-run elasticities. As for elasticities of substitution between other factors, our results are close to Griffin and Gregory for labor and energy, but we find greater substitution of capital and labor.

There are two ways to reconcile the capital-energy substitutability found by this study and also observed by Griffin and Gregory with the complementarity observed by Berndt and Wood. First, as mentioned above, the use of international data is more likely to permit estimation of the long-run cost function, and we might expect to observe complementarity in the shorter run and substitutability in the long run. Second, we must keep in mind that both Griffin and Gregory and this study use a three-factor total cost function, while Berndt and Wood work with four factors. As recently pointed out by Berndt and Wood (23), complementarity between two factors in a four-dimensional production space can be consistent with substitutability between the same factors in a three-dimensional subspace. It might be, then, that were we to have included nonenergy raw materials as a fourth factor of production, we would have observed energy-capital complementarity. Unfortunately the data were not available to test this hypothesis.

We must also recognize that the answer to the capital-energy substitutability question may be submerged in the high level of aggregation of this study as well as the Griffin-Gregory and Berndt-Wood studies. As can be seen in table 5.14, Halvorsen and Ford (79) found considerable variation across industries in the elasticity of substitution for capital and energy (as well as in other elasticities of substitution), and their findings are supported by the work of Moroney and Toevs (119, 120), which also shows variation in factor substitutability (capital, labor, and all raw materials) across industries. We should in fact expect factor substitutability to differ considerably across individual industries; the question is whether this variation might lead to aggregation bias in studies such as this one. We have not been able to test for such bias, so we must be aware of the possibility of its presence.

Let us now turn to the own-price elasticity of aggregate energy use. This is an important parameter for any energy policy debate, and provides an example of where the conventional wisdom may have been very misleading. Our estimate (about −0.8), together with that of Griffin and Gregory (also −0.8), is larger than most other estimates, which fall in the range of −0.3 to −0.6, and is larger still than the consensus estimates often used for policy analysis in the U.S. (−0.2 to −0.3). Again, most other estimates are based on time-series data for a single country, and may be short run.

It is more difficult to find a consensus on partial and total fuel-price elasticities. Although most would agree that electricity demand is less elastic than the demands for other fuels, partial long-run elasticities for the U.S. range from −0.5 to −1.2. Our study finds electricity demand to be even less elastic; we found partial own-price elasticities to range from −0.07 to −0.16. Our total own-price elasticity estimates, however, are closer to the estimates of others (largely because of our higher estimate of the own-price elasticity of energy). We find this elasticity to range from −0.54 to −0.63, where Halvorsen (77) obtained an estimate of −0.92 and Fuss (60) −0.74. Our own-price elasticity estimate (total) for oil is also well below the estimates of others; −0.22 to −1.17 as compared to Halvorsen's estimate of −2.82 and Fuss's of −1.30. An explanation for this discrepancy will probably require further work. There is less disagreement over the elasticities for coal and natural gas. Our estimates of the total own-price elasticities for coal (−1.29 to −2.24) and natural gas (−0.41 to −2.34) are generally in line with other estimates.

We have obtained elasticity estimates in this chapter that we expect to pertain to the long run. Unfortunately, our results do not tell us just how long the long run is and what the speeds of adjustment of the various elasticity estimates are. Depreciation rate data for industrial fuel-burning equipment show considerable variation across industries, so that the median lag (or half-life) for energy demand elasticities might vary from three or four years to fifteen years across different industries. As a rough guess, however, we might expect that on average the median lag for the own-price elasticity of aggregate industrial energy demand is around ten years. An accurate determination of this lag would of course require the estimation of explicitly dynamic demand models, such as the dynamic version of the translog cost function outlined in chapter 2 of this book.

Although the response time may be considerable, our results indicate that price elasticities of industrial energy demand are much larger in the long run than had been thought previously. This means that price-oriented policies to reduce industrial energy use can be very effective, given that enough time is allowed to pass to allow the price mechanism to work. In the past some countries have been reluctant to use such policies, perhaps because of a misguided use of low consensus estimates of demand elasticities. Hopefully policies adopted in the future will be based on the recognition that these elasticities can be large in the long run.

## Appendix: Logit Models of Industrial Fuel Shares

We have also estimated a number of static and dynamic logit models to describe the dependence of fuel shares on prices and the level of output. The decision functions in these models are linear or logarithmic functions of relative fuel prices $\tilde{P}_i$ (the price of fuel $i$ divided by the price of energy), the level of output, and, in the case of the dynamic models, lagged shares. Recall from equations (2.63) and (2.64) that this leads to a set of three equations that must be estimated simultaneously, since certain coefficients are constrained to be the same across equations.[6] We therefore use iterative Zellner estimation to estimate all of the models.

Our static logit models are of the general form

$$\log (S_i/S_4) = \sum_{k=1}^{10} a_{i4k}D_k + b_i\tilde{P}_i - b_4\tilde{P}_4 + c_{i4}Q, \quad i = 1, 2, 3, \tag{5.4}$$

where $a_{i4k} = a_{ik} - a_{4k}$, and $c_{i4} = c_i - c_4$, as in equation (2.64). The $D_k$ are country dummy variables, and the fuels are ordered (1) solid, (2) liquid, (3) gas, and (4) electricity.

The dynamic models are based on the assumption that the choice of fuels this period depends on the relative shares last period, as well as this period's relative prices and output. The dependence on past shares is intended to incorporate partial adjustment of the capital stock. It leads to equations of the form

$$\log (S_i/S_4) = \sum_{k=1}^{10} a_{i4k}D_k + b_i\tilde{P}_i - b_4\tilde{P}_4 + c_{i4}Q + d_iS_{i,t-1} - d_4S_{4,t-1}, \quad i = 1, 2, 3. \tag{5.5}$$

6. Even without cross-equation coefficient constraints, simultaneous equation estimation is desirable in that insofar as errors are correlated across equations, it yields more efficient parameter estimates.

Note that this is not a Koyck adjustment model. The coefficients $d_i$ can be greater than 1 (although we would expect them to be positive), and in general a change in price will not lead to geometrically declining changes in shares over time.

As was the case for the residential sector, the results of estimating various versions of these models are unfortunately disappointing, and are not reported in detail here. In all of the models that we have estimated the own-price elasticities are positive for both solid and liquid fuel. In addition the own-price elasticities for gas and electricity are insignificant or positive for some of the models. The logit models provides essentially an ad hoc and somewhat restrictive description of fuel choice by industrial consumers of energy, and that description is not consistent with the data and our a priori expectations regarding the characteristics of fuel demands.

**Table 5.1**
Parameter estimates for fuel choice models

| Parameter[a] | (1) Ten countries, regionally homogeneous, 1959–1973 | (2) U.S. and Canada, 1959–1973 | (3) U.S. and Canada, 1959–1974 | (4) Europe and Japan, 1959–1973 |
|---|---|---|---|---|
| $\alpha_1$ | 0.1038 | | | |
| | (0.0242) | | | |
| $\alpha_1 D_1$ | | 0.2335 | 0.2092 | 0.2438 |
| | | (0.0218) | (0.0186) | (0.0268) |
| $\alpha_1 D_2$ | | 0.2555 | 0.1950 | 0.2379 |
| | | (0.0382) | (0.0331) | (0.0231) |
| $\alpha_1 D_3$ | | | | 0.2571) |
| | | | | (0.0314) |
| $\alpha_1 D_4$ | | | | 0.1486 |
| | | | | (0.0368) |
| $\alpha_1 D_5$ | | | | −0.0613 |
| | | | | (0.0213) |
| $\alpha_1 D_6$ | | | | 0.1221 |
| | | | | (0.0280) |
| $\alpha_1 D_7$ | | | | 0.1768 |
| | | | | (0.0346) |
| $\alpha_1 D_8$ | | | | 0.2864 |
| | | | | (0.0286) |
| $\alpha_1 D_9$ | | | | |
| $\alpha_1 D_{10}$ | | | | |
| $\alpha_2$ | 0.3045 | | | |
| | (0.0136) | | | |
| $\alpha_2 D_1$ | | 0.2099 | 0.2385 | 0.3807 |
| | | (0.0162) | (0.0132) | (0.0228) |
| $\alpha_2 D_2$ | | 0.0501 | 0.1265 | 0.4052 |
| | | (0.0305) | (0.0253) | (0.0212) |
| $\alpha_2 D_3$ | | | | 0.3573 |
| | | | | (0.0255) |

[a] 1 = solid, 2 = liquid, 3 = gas, 4 = electricity. The countries are numbered alphabetically: 1 = Belgium, 2 = Canada, 3 = France, 4 = Italy, 5 = Netherlands, 6 = Norway, 7 = Sweden, 8 = U.K., 9 = U.S., 10 = West Germany.

| (5) Europe and Japan, 1959-1974 | (6) Ten countries, 1959-1973 | (7) Ten countries, 1959-1974 | (8) Europe and Japan 1961-1973, with three-year intervals |
|---|---|---|---|
| 0.2560 | 0.1836 | 0.2028 | 0.2472 |
| (0.0243) | (0.0262) | (0.0240) | (0.0332) |
| 0.2503 | 0.2763 | 0.2890 | 0.2524 |
| (0.0210) | (0.0242) | (0.0220) | (0.0277) |
| 0.2728 | 0.2780 | 0.2950 | 0.2907 |
| (0.0287) | (0.0220) | (0.0204) | (0.0378) |
| 0.1686 | 0.2959 | 0.3167 | 0.1702 |
| (0.0330) | (0.0280) | (0.0261) | (0.0417) |
| −0.0504 | 0.1991 | 0.2226 | −0.0640 |
| (0.0197) | (0.0332) | (0.0303) | (0.0298) |
| 0.1418 | −0.0575 | −0.0518 | 0.1291 |
| (0.0251) | (0.0188) | (0.0174) | (0.0348) |
| 0.1961 | 0.1504 | 0.1688 | 0.1879 |
| (0.0316) | (0.0249) | (0.0223) | (0.0413) |
| 0.2963 | 0.2088 | 0.2282 | 0.2881 |
| (0.0259) | (0.0303) | (0.0278) | (0.0331) |
| | 0.1531 | 0.1815 | |
| | (0.0358) | (0.0330) | |
| | 0.3289 | 0.3422 | |
| | (0.0263) | (0.0242) | |
| | | | |
| 0.4107 | 0.3082 | 0.3143 | 0.4046 |
| (0.0204) | (0.0209) | (0.0188) | (0.0337) |
| 0.4371 | 0.3477 | 0.3761 | 0.4408 |
| (0.0188) | (0.0213) | (0.0186) | (0.0278) |
| 0.3843 | 0.3636 | 0.3895 | 0.3900 |
| (0.0228) | (0.0207) | (0.0180) | (0.0341) |

**Table 5.1**
(continued)

| Parameter[a] | (1) Ten countries, regionally homogeneous, 1959–1973 | (2) U.S. and Canada, 1959–1973 | (3) U.S. and Canada, 1959–1974 | (4) Europe and Japan, 1959–1973 |
|---|---|---|---|---|
| $\alpha_2 D_4$ | | | | 0.4137 |
| | | | | (0.0301) |
| $\alpha_2 D_5$ | | | | 0.3317 |
| | | | | (0.0178) |
| $\alpha_2 D_6$ | | | | 0.3839 |
| | | | | (0.0226) |
| $\alpha_2 D_7$ | | | | 0.4037 |
| | | | | (0.0261) |
| $\alpha_2 D_8$ | | | | 0.3513 |
| | | | | (0.0244) |
| $\alpha_2 D_9$ | | | | |
| $\alpha_2 D_{10}$ | | | | |
| $\alpha_3$ | 0.0544 | | | |
| | (0.0122) | | | |
| $\alpha_3 D_1$ | | 0.2997 | 0.2944 | 0.1934 |
| | | (0.0220) | (0.0210) | (0.0147) |
| $\alpha_3 D_2$ | | 0.4705 | 0.4680 | 0.1466 |
| | | (0.0325) | (0.0298) | (0.0125) |
| $\alpha_3 D_3$ | | | | 0.1638 |
| | | | | (0.0153) |
| $\alpha_3 D_4$ | | | | 0.2311 |
| | | | | (0.0179) |
| $\alpha_3 D_5$ | | | | 0.0888 |
| | | | | (0.0140) |
| $\alpha_3 D_6$ | | | | 0.1362 |
| | | | | (0.0151) |
| $\alpha_3 D_7$ | | | | 0.2362 |
| | | | | (0.0171) |
| $\alpha_3 D_8$ | | | | 0.1356 |
| | | | | (0.0145) |
| $\alpha_3 D_9$ | | | | |

| (5) Europe and Japan, 1959–1974 | (6) Ten countries, 1959–1973 | (7) Ten countries, 1959–1974 | (8) Europe and Japan 1961–1973, with three-year intervals |
|---|---|---|---|
| 0.4456 | 0.3240 | 0.3433 | 0.4388 |
| (0.0262) | (0.0234) | (0.0205) | (0.0375) |
| 0.3422 | 0.3657 | 0.3918 | 0.3287 |
| (0.0166) | (0.0284) | (0.0241) | (0.0308) |
| 0.4113 | 0.3277 | 0.3438 | 0.4000 |
| (0.0199) | (0.0158) | (0.0146) | (0.0319) |
| 0.4311 | 0.3567 | 0.3843 | 0.4212 |
| (0.0234) | (0.0238) | (0.0179) | (0.0361) |
| 0.3858 | 0.3757 | 0.4013 | 0.3891 |
| (0.0214) | (0.0238) | (0.0208) | (0.0312) |
| | 0.2292 | 0.2425 | |
| | (0.0274) | (0.0240) | |
| | 0.3096 | 0.3390 | |
| | (0.0233) | (0.0200) | |
| | | | |
| 0.1901 | 0.2066 | 0.2069 | 0.2237 |
| (0.0141) | (0.0132) | (0.0126) | (0.0247) |
| 0.1432 | 0.1991 | 0.1933 | 0.1554 |
| (0.0120) | (0.0134) | (0.0132) | (0.0186) |
| 0.1640 | 0.1555 | 0.1484 | 0.1733 |
| (0.0146) | (0.0129) | (0.0126) | (0.0228) |
| 0.2366 | 0.1771 | 0.1758 | 0.2749 |
| (0.0167) | (0.0151) | (0.0148) | (0.0256) |
| 0.0843 | 0.2438 | 0.2462 | 0.1056 |
| (0.0133) | (0.0176) | (0.0169) | (0.0257) |
| 0.1301 | 0.0872 | 0.0826 | 0.1611 |
| (0.0141) | (0.0128) | (0.0121) | (0.0243) |
| 0.2336 | 0.1418 | 0.1341 | 0.2684 |
| (0.0161) | (0.0140) | (0.0131) | (0.0272) |
| 0.1371 | 0.2445 | 0.2411 | 0.1585 |
| (0.0138) | (0.0160) | (0.0151) | (0.0211) |
| | 0.3756 | 0.3732 | |
| | (0.0172) | (0.0164) | |

**Table 5.1**
(continued)

| Parameter[a] | (1) Ten countries, regionally homogeneous, 1959–1973 | (2) U.S. and Canada, 1959–1973 | (3) U.S. and Canada, 1959–1974 | (4) Europe and Japan, 1959–1973 |
|---|---|---|---|---|
| $\alpha_3 D_{10}$ | | | | |
| $\alpha_4$ | 0.5374 | | | |
| | (0.0222) | | | |
| $\alpha_4 D_1$ | | 0.2569 | 0.2579 | 0.1821 |
| | | (0.0233) | (0.0191) | (0.0223) |
| $\alpha_4 D_2$ | | 0.2238 | 0.2105 | 0.2104 |
| | | (0.0386) | (0.0300) | (0.0226) |
| $\alpha_4 D_3$ | | | | 0.2218 |
| | | | | (0.0271) |
| $\alpha_4 D_4$ | | | | 0.2067 |
| | | | | (0.0334) |
| $\alpha_4 D_5$ | | | | 0.6408 |
| | | | | (0.0116) |
| $\alpha_4 D_6$ | | | | 0.3577 |
| | | | | (0.0116) |
| $\alpha_4 D_7$ | | | | 0.1833 |
| | | | | (0.0266) |
| $\alpha_4 D_8$ | | | | 0.2268 |
| | | | | (0.0266) |
| $\alpha_4 D_9$ | | | | |
| $\alpha_4 D_{10}$ | | | | |
| $\beta_{11}$ | −0.0723 | −0.1017 | −0.0842 | −0.1039 |
| | (0.0196) | (0.0406) | (0.0352) | (0.0293) |
| $\beta_{12}$ | 0.0195 | 0.0739 | 0.0215 | 0.0093 |
| | (0.0109) | (0.0256) | (0.0211) | (0.0210) |
| $\beta_{13}$ | 0.0252 | 0.1195 | 0.1203 | 0.1184 |
| | (0.0084) | (0.0201) | (0.0183) | (0.0112) |
| $\beta_{14}$ | 0.0276 | −0.0917 | −0.0576 | −0.0237 |
| | (0.0124) | (0.0164) | (0.0142) | (0.0142) |
| $\beta_{21}$ | 0.0195 | 0.0739 | 0.0215 | 0.0093 |
| | (0.0109) | (0.0256) | (0.0211) | (0.0210) |

| (5) Europe and Japan, 1959–1974 | (6) Ten countries, 1959–1973 | (7) Ten countries, 1959–1974 | (8) Europe and Japan 1961–1973, with three-year intervals |
|---|---|---|---|
| | 0.1458 | 0.1443 | |
| | (0.0144) | (0.0140) | |
| | | | |
| 0.1432 | 0.3015 | 0.2760 | 0.1245 |
| (0.0183) | (0.0196) | (0.0174) | (0.0254) |
| 0.1694 | 0.1769 | 0.1416 | 0.1514 |
| (0.0184) | (0.0220) | (0.0180) | (0.0258) |
| 0.1789 | 0.2029 | 0.1617 | 0.1460 |
| (0.0229) | (0.0229) | (0.0190) | (0.0319) |
| 0.1492 | 0.2030 | 0.1642 | 0.1161 |
| (0.0273) | (0.0270) | (0.0233) | (0.0374) |
| 0.6239 | 0.1915 | 0.1395 | 0.6297 |
| (0.0101) | (0.0332) | (0.0275) | (0.0136) |
| 0.3168 | 0.6426 | 0.6254 | 0.3098 |
| (0.0176) | (0.0110) | (0.0097) | (0.0244) |
| 0.1392 | 0.3511 | 0.3128 | 0.1225 |
| (0.0221) | (0.0213) | (0.0172) | (0.0300) |
| 0.1808 | 0.1710 | 0.1294 | 0.1643 |
| (0.0219) | (0.0258) | (0.0216) | (0.0302) |
| | 0.2421 | 0.2027 | |
| | (0.0301) | (0.0260) | |
| | 0.2158 | 0.1745 | |
| | (0.0267) | (0.0221) | |
| −0.0826 | −0.1103 | −0.0910 | −0.0792 |
| (0.0272) | (0.0243) | (0.0224) | (0.0384) |
| −0.0001 | 0.0280 | 0.0184 | −0.0138 |
| (0.0196) | (0.0176) | (0.0160) | (0.0302) |
| 0.1150 | 0.1238 | 0.1243 | 0.1320 |
| (0.0103) | (0.0100) | (0.0092) | (0.0172) |
| −0.0323 | −0.0415 | −0.0517 | −0.0391 |
| (0.0128) | (0.0126) | (0.0115) | (0.0173) |
| −0.0001 | 0.0280 | 0.0184 | −0.0138 |
| (0.0196) | (0.0176) | (0.0160) | (0.0302) |

**Table 5.1**
(continued)

| Parameter[a] | (1) Ten countries, regionally homogeneous, 1959–1973 | (2) U.S. and Canada, 1959–1973 | (3) U.S. and Canada, 1959–1974 | (4) Europe and Japan,. 1959–1973 |
|---|---|---|---|---|
| $\beta_{22}$ | 0.0445 | −0.0152 | 0.0801 | 0.1063 |
| | (0.0099) | (0.0233) | (0.0216) | (0.0210) |
| $\beta_{23}$ | 0.0024 | −0.0754 | −0.0686 | −0.0240 |
| | (0.0057) | (0.0128) | (0.0100) | (0.0105) |
| $\beta_{24}$ | −0.0664 | 0.0167 | −0.0330 | −0.0916 |
| | (0.0069) | (0.0151) | (0.0135) | (0.0110) |
| $\beta_{31}$ | 0.0252 | 0.1195 | 0.1203 | 0.1184 |
| | (0.0084) | (0.0201) | (0.0183) | (0.0112) |
| $\beta_{32}$ | 0.0024 | −0.0754 | −0.0686 | −0.0240 |
| | (0.0057) | (0.0128) | (0.0100) | (0.0105) |
| $\beta_{33}$ | −0.0429 | 0.0557 | 0.0477 | −0.0340 |
| | (0.0064) | (0.0177) | (0.0171) | (0.0087) |
| $\beta_{34}$ | 0.0153 | −0.0998 | −0.0995 | −0.0604 |
| | (0.0061) | (0.0141) | (0.0127) | (0.0063) |
| $\beta_{41}$ | 0.0276 | −0.0917 | −0.0576 | −0.0237 |
| | (0.0124) | (0.0164) | (0.0142) | (0.0142) |
| $\beta_{42}$ | −0.0664 | 0.0167 | −0.0330 | −0.0916 |
| | (0.0069) | (0.0151) | (0.0134) | (0.0110) |
| $\beta_{43}$ | 0.0153 | 0.0998 | −0.0995 | −0.0604 |
| | (0.0061) | (0.0141) | (0.0127) | (0.0063) |
| $\beta_{44}$ | 0.0234 | 0.1748 | 0.1902 | 0.1758 |
| | (0.0116) | (0.0176) | (0.0139) | (0.0126) |
| RSQ | | | | |
| Equation 1 | 0.045 | 0.260 | 0.316 | 0.746 |
| Equation 2 | 0.400 | 0.959 | 0.928 | 0.471 |
| Equation 3 | 0.227 | 0.955 | 0.954 | 0.633 |

| (5) Europe and Japan, 1959–1974 | (6) Ten countries, 1959–1973 | (7) Ten countries, 1959–1974 | (8) Europe and Japan 1961–1973, with three-year intervals |
|---|---|---|---|
| 0.1282 | 0.0762 | 0.1000 | 0.1321 |
| (0.0194) | (0.0182) | (0.0161) | (0.0336) |
| −0.0234 | −0.0279 | −0.0318 | −0.0171 |
| (0.0098) | (0.0092) | (0.0084) | (0.0192) |
| −0.1047 | −0.0763 | −0.0865 | −0.1012 |
| (0.0095) | (0.0101) | (0.0085) | (0.0132) |
| 0.1150 | 0.1238 | 0.1243 | 0.1320 |
| (0.0103) | (0.0100) | (0.0092) | (0.0172) |
| −0.0234 | −0.0279 | −0.0318 | −0.0171 |
| (0.0098) | (0.0092) | (0.0084) | (0.0192) |
| −0.0316 | −0.0299 | −0.0275 | −0.0451 |
| (0.0081) | (0.0080) | (0.0075) | (0.0163) |
| −0.0998 | −0.0661 | −0.0650 | −0.0697 |
| (0.0141) | (0.0063) | (0.0060) | (0.0084) |
| −0.0323 | −0.0415 | −0.0517 | −0.0391 |
| (0.0128) | (0.0126) | (0.0115) | (0.0173) |
| −0.1047 | −0.0763 | −0.0865 | −0.1012 |
| (0.0095) | (0.0101) | (0.0085) | (0.0132) |
| −0.0998 | −0.0661 | −0.0650 | −0.0697 |
| (0.0141) | (0.0063) | (0.0060) | (0.0084) |
| 0.1969 | 0.1839 | 0.2031 | 0.2100 |
| (0.0105) | (0.0124) | (0.0105) | (0.0143) |
| | | | |
| 0.733 | 0.777 | 0.769 | 0.799 |
| 0.538 | 0.685 | 0.705 | 0.524 |
| 0.617 | 0.824 | 0.802 | 0.681 |

**Table 5.2**
Partial Fuel-Price Elasticities (U.S. and Canada estimated separately, country dummy variables)

| Elasticity[a] | Version[b] | Canada | France | Italy | Japan |
|---|---|---|---|---|---|
| $\eta_{11}$ | (a) | −1.80 | −1.04 | −1.49 | −1.32 |
| | | (0.36) | (0.10) | (0.18) | (0.15) |
| | (b) | −1.64 | −0.97 | −1.36 | −1.21 |
| | | (0.32) | (0.09) | (0.17) | (0.13) |
| $\eta_{12}$ | (a) | 0.91 | 0.20 | 0.27 | 0.21 |
| | | (0.23) | (0.07) | (0.13) | (0.11) |
| | (b) | 0.44 | 0.17 | 0.21 | 0.17 |
| | | (0.19) | (0.06) | (0.12) | (0.10) |
| $\eta_{13}$ | (a) | 1.17 | 0.46 | 0.83 | 0.65 |
| | | (0.18) | (0.04) | (0.07) | (0.06) |
| | (b) | 1.17 | 0.44 | 0.81 | 0.64 |
| | | (0.16) | (0.03) | (0.06) | (0.05) |
| $\eta_{14}$ | (a) | −0.28 | 0.39 | 0.39 | 0.45 |
| | | (0.15) | (0.05) | (0.09) | (0.07) |
| | (b) | 0.03 | 0.36 | 0.34 | 0.41 |
| | | (0.13) | (0.04) | (0.08) | (0.06) |
| $\eta_{21}$ | (a) | 0.41 | 0.36 | 0.20 | 0.26 |
| | | (0.10) | (0.12) | (0.10) | (0.13) |
| | (b) | 0.20 | 0.30 | 0.16 | 0.20 |
| | | (0.08) | (0.12) | (0.10) | (0.12) |
| $\eta_{22}$ | (a) | −0.81 | −0.20 | −0.29 | −0.20 |
| | | (0.09) | (0.12) | (0.10) | (0.13) |
| | (b) | −0.42 | −0.07 | −0.18 | −0.06 |
| | | (0.09) | (0.11) | (0.10) | (0.12) |
| $\eta_{23}$ | (a) | −0.21 | −0.08 | −0.03 | −0.08 |
| | | (0.05) | (0.06) | (0.05) | (0.06) |
| | (b) | −0.18 | −0.07 | −0.02 | −0.08 |
| | | (0.04) | (0.06) | (0.05) | (0.06) |
| $\eta_{24}$ | (a) | 0.61 | −0.08 | 0.11 | 0.02 |
| | | (0.06) | (0.07) | (0.05) | (0.07) |
| | (b) | 0.41 | −0.15 | 0.04 | −0.05 |
| | | (0.05) | (0.06) | (0.05) | (0.06) |
| $\eta_{31}$ | (a) | 1.35 | 2.21 | 1.52 | 2.12 |
| | | (0.21) | (0.18) | (0.13) | (0.18) |

[a] 1 = solid, 2 = liquid, 3 = gas, 4 = electricity.
[b] Model version (a) 1959–1973, (b) 1959–1974.
[c] Almost no natural gas is consumed in the industrial sectors of Norway and Sweden, so that these elasticities are meaningless.

| Netherlands | Norway | Sweden | U.K. | U.S. | West Germany |
|---|---|---|---|---|---|
| −1.67 | −2.08 | −1.26 | −1.12 | −2.17 | −1.09 |
| (0.22) | (0.33) | (0.13) | (0.11) | (0.50) | (0.10) |
| −1.50 | −1.84 | −1.15 | −1.04 | −1.95 | −0.99 |
| (0.21) | (0.31) | (0.12) | (0.10) | (0.43) | (0.09) |
| 0.21 | 0.37 | 0.24 | 0.21 | 0.99 | 0.15 |
| (0.16) | (0.24) | (0.10) | (0.08) | (0.31) | (0.07) |
| 0.14 | 0.27 | 0.20 | 0.18 | 0.34 | 0.12 |
| (0.15) | (0.22) | (0.09) | (0.07) | (0.26) | (0.07) |
| 0.98 | 1.34 | 0.55 | 0.52 | 1.66 | 0.43 |
| (0.09) | (0.13) | (0.05) | (0.04) | (0.25) | (0.04) |
| 0.95 | 1.30 | 0.53 | 0.50 | 1.66 | 0.42 |
| (0.08) | (0.12) | (0.05) | (0.04) | (0.22) | (0.04) |
| 0.48 | 0.37 | 0.46 | 0.39 | −0.48 | 0.49 |
| (0.11) | (0.16) | (0.07) | (0.05) | (0.20) | (0.05) |
| 0.41 | 0.28 | 0.42 | 0.36 | −0.06 | 0.46 |
| (0.10) | (0.14) | (0.06) | (0.05) | (0.17) | (0.04) |
| 0.20 | 0.12 | 0.27 | 0.32 | 0.97 | 0.37 |
| (0.15) | (0.08) | (0.10) | (0.12) | (0.31) | (0.18) |
| 0.13 | 0.09 | 0.22 | 0.27 | 0.34 | 0.29 |
| (0.14) | (0.07) | (0.10) | (0.11) | (0.25) | (0.17) |
| −0.11 | −0.34 | −0.27 | −0.22 | −1.10 | 0.03 |
| (0.15) | (0.08) | (0.10) | (0.12) | (0.28) | (0.18) |
| 0.04 | −0.25 | −0.16 | −0.09 | 0.05 | 0.22 |
| (0.14) | (0.07) | (0.10) | (0.11) | (0.25) | (0.17) |
| −0.10 | −0.09 | −0.11 | −0.06 | −0.72 | −0.18 |
| (0.07) | (0.04) | (0.05) | (0.06) | (0.15) | (0.09) |
| −0.10 | −0.09 | −0.11 | −0.06 | −0.64 | −0.18 |
| (0.07) | (0.03) | (0.05) | (0.06) | (0.12) | (0.08) |
| 0.01 | 0.30 | 0.12 | −0.04 | 0.85 | −0.22 |
| (0.08) | (0.04) | (0.05) | (0.06) | (0.18) | (0.10) |
| −0.08 | 0.25 | 0.05 | −0.11 | 0.24 | −0.34 |
| (0.07) | (0.04) | (0.05) | (0.05) | (0.16) | (0.08) |
| 1.84 |  |  | 1.86 | 0.72 | 4.98 |
| (0.16) | c | c | (0.15) | (0.11) | (0.44) |

**Table 5.2**
(continued)

| Elasticity[a] | Version[b] | Canada | France | Italy | Japan |
|---|---|---|---|---|---|
| | (b) | 1.36 | 2.16 | 1.48 | 2.06 |
| | | (0.19) | (0.17) | (0.12) | (0.17) |
| $\eta_{32}$ | (a) | −0.53 | −0.22 | −0.06 | −0.22 |
| | | (0.13) | (0.17) | (0.12) | (0.17) |
| | (b) | −0.45 | −0.21 | −0.06 | −0.21 |
| | | (0.10) | (0.16) | (0.11) | (0.16) |
| $\eta_{33}$ | (a) | −0.33 | −1.49 | −1.30 | −1.49 |
| | | (0.18) | (0.14) | (0.10) | (0.14) |
| | (b) | −0.41 | −1.45 | −1.28 | −1.45 |
| | | (0.18) | (0.13) | (0.09) | (0.13) |
| $\eta_{34}$ | (a) | −0.49 | −0.51 | −0.15 | −0.41 |
| | | (0.15) | (0.10) | (0.07) | (0.10) |
| | (b) | −0.49 | −0.50 | −0.15 | −0.40 |
| | | (0.13) | (0.09) | (0.07) | (0.09) |
| $\eta_{41}$ | (a) | −0.06 | 0.25 | 0.12 | 0.16 |
| | | (0.03) | (0.03) | (0.03) | (0.02) |
| | (b) | 0.01 | 0.23 | 0.10 | 0.14 |
| | | (0.03) | (0.03) | (0.02) | (0.02) |
| $\eta_{42}$ | (a) | 0.28 | −0.03 | 0.04 | 0.01 |
| | | (0.03) | (0.02) | (0.02) | (0.02) |
| | (b) | 0.19 | −0.05 | 0.02 | −0.02 |
| | | (0.02) | (0.02) | (0.02) | (0.02) |
| $\eta_{43}$ | (a) | −0.09 | −0.07 | −0.02 | −0.04 |
| | | (0.03) | (0.01) | (0.01) | (0.01) |
| | (b) | −0.09 | −0.07 | −0.02 | −0.04 |
| | | (0.02) | (0.01) | (0.01) | (0.01) |
| $\eta_{44}$ | (a) | −0.14 | −0.16 | −0.13 | −0.12 |
| | | (0.03) | (0.03) | (0.02) | (0.02) |
| | (b) | −0.11 | −0.11 | −0.09 | −0.08 |
| | | (0.03) | (0.02) | (0.02) | (0.02) |

| Netherlands | Norway | Sweden | U.K. | U.S. | West Germany |
|---|---|---|---|---|---|
| 1.80 | | | 1.82 | 0.72 | 4.84 |
| (0.15) | [c] | [c] | (0.14) | (0.10) | (0.41) |
| −0.21 | | | −0.15 | −0.32 | −0.83 |
| (0.15) | [c] | [c] | (0.14) | (0.07) | (0.42) |
| −0.20 | | | −0.14 | −0.28 | −0.81 |
| (0.14) | [c] | [c] | (0.13) | (0.05) | (0.39) |
| −1.42 | | | −1.38 | −0.52 | −2.31 |
| (0.13) | [c] | [c] | (0.12) | (0.09) | (0.34) |
| −1.39 | | | −1.35 | −0.55 | −2.23 |
| (0.12) | [c] | [c] | (0.11) | (0.09) | (0.32) |
| −0.22 | | | −0.33 | 0.12 | −1.82 |
| (0.09) | [c] | [c] | (0.09) | (0.08) | (0.25) |
| −0.21 | | | −0.33 | 0.12 | −1.81 |
| (0.08) | [c] | [c] | (0.08) | (0.07) | (0.23) |
| 0.09 | 0.05 | 0.18 | 0.22 | −0.06 | 0.25 |
| (0.02) | (0.02) | (0.02) | (0.03) | (0.03) | (0.03) |
| 0.08 | 0.03 | 0.16 | 0.20 | −0.01 | 0.23 |
| (0.02) | (0.02) | (0.02) | (0.03) | (0.02) | (0.02) |
| 0.00 | 0.12 | 0.04 | −0.01 | 0.11 | −0.05 |
| (0.02) | (0.02) | (0.02) | (0.02) | (0.02) | (0.02) |
| −0.02 | 0.10 | 0.02 | −0.04 | 0.03 | −0.07 |
| (0.01) | (0.01) | (0.01) | (0.02) | (0.02) | (0.02) |
| −0.02 | −0.09 | −0.10 | −0.05 | 0.03 | −0.08 |
| (0.01) | (0.01) | (0.01) | (0.01) | (0.02) | (0.01) |
| −0.02 | −0.09 | −0.10 | −0.05 | 0.03 | −0.08 |
| (0.01) | (0.01) | (0.01) | (0.01) | (0.02) | (0.01) |
| −0.07 | −0.08 | −0.12 | −0.15 | −0.08 | −0.12 |
| (0.02) | (0.02) | (0.02) | (0.03) | (0.03) | (0.02) |
| −0.04 | −0.05 | −0.08 | −0.11 | −0.06 | −0.08 |
| (0.02) | (0.02) | (0.02) | (0.02) | (0.02) | (0.02) |

**Table 5.3**
Partial Fuel-Price Elasticities (ten countries pooled, country dummy variables)

| Elasticity[a] | Version[b] | Canada | France | Italy | Japan |
|---|---|---|---|---|---|
| $\eta_{11}$ | (a) | −1.87 | −1.07 | −1.53 | −1.35 |
| | | (0.22) | (0.08) | (0.15) | (0.12) |
| | (b) | −1.70 | −1.00 | −1.41 | −1.25 |
| | | (0.20) | (0.07) | (0.14) | (0.11) |
| $\eta_{12}$ | (a) | 0.50 | 0.26 | 0.39 | 0.31 |
| | | (0.16) | (0.06) | (0.11) | (0.09) |
| | (b) | 0.41 | 0.23 | 0.33 | 0.26 |
| | | (0.14) | (0.05) | (0.10) | (0.08) |
| $\eta_{13}$ | (a) | 1.20 | 0.47 | 0.86 | 0.68 |
| | | (0.09) | (0.03) | (0.06) | (0.05) |
| | (b) | 1.21 | 0.47 | 0.87 | 0.68 |
| | | (0.08) | (0.03) | (0.06) | (0.05) |
| $\eta_{14}$ | (a) | 0.17 | 0.33 | 0.28 | 0.36 |
| | | (0.11) | (0.04) | (0.08) | (0.06) |
| | (b) | 0.08 | 0.30 | 0.22 | 0.31 |
| | | (0.10) | (0.04) | (0.07) | (0.06) |
| $\eta_{21}$ | (a) | 0.22 | 0.47 | 0.29 | 0.37 |
| | | (0.07) | (0.10) | (0.08) | (0.11) |
| | (b) | 0.18 | 0.41 | 0.25 | 0.31 |
| | | (0.06) | (0.09) | (0.08) | (0.10) |
| $\eta_{22}$ | (a) | −0.44 | −0.38 | −0.43 | −0.38 |
| | | (0.07) | (0.10) | (0.09) | (0.11) |
| | (b) | −0.35 | −0.24 | −0.31 | −0.23 |
| | | (0.06) | (0.10) | (0.08) | (0.10) |
| $\eta_{23}$ | (a) | −0.02 | −0.10 | −0.05 | −0.11 |
| | | (0.04) | (0.05) | (0.04) | (0.06) |
| | (b) | −0.03 | −0.13 | −0.06 | −0.13 |
| | | (0.03) | (0.05) | (0.04) | (0.05) |
| $\eta_{24}$ | (a) | 0.24 | 0.02 | 0.18 | 0.11 |
| | | (0.04) | (0.06) | (0.05) | (0.06) |
| | (b) | 0.19 | −0.04 | 0.13 | 0.05 |
| | | (0.03) | (0.05) | (0.04) | (0.05) |
| $\eta_{31}$ | (a) | 1.39 | 2.30 | 1.58 | 2.21 |
| | | (0.10) | (0.16) | (0.11) | (0.16) |

[a] 1 = solid, 2 = liquid, 3 = gas, 4 = electricity.
[b] Model version (a) 1959–1973, (b) 1959–1974.
[c] Almost no natural gas is consumed in the industrial sectors of Norway and Sweden, so that these elasticities are meaningless.

| Netherlands | Norway | Sweden | U.K. | U.S. | West Germany |
|---|---|---|---|---|---|
| −1.71 | −2.15 | −1.29 | −1.14 | −2.27 | −1.10 |
| (0.19) | (0.27) | (0.11) | (0.09) | (0.30) | (0.08) |
| −1.56 | −1.93 | −1.20 | −1.07 | −2.04 | −1.02 |
| (0.17) | (0.25) | (0.10) | (0.08) | (0.28) | (0.08) |
| 0.36 | 0.58 | 0.32 | 0.28 | 0.43 | 0.21 |
| (0.13) | (0.20) | (0.08) | (0.07) | (0.22) | (0.06) |
| 0.28 | 0.47 | 0.29 | 0.24 | 0.31 | 0.18 |
| (0.12) | (0.18) | (0.07) | (0.06) | (0.20) | (0.06) |
| 1.02 | 1.40 | 0.58 | 0.54 | 1.71 | 0.45 |
| (0.08) | (0.11) | (0.05) | (0.04) | (0.12) | (0.03) |
| 1.02 | 1.40 | 0.58 | 0.54 | 1.71 | 0.45 |
| (0.07) | (0.10) | (0.04) | (0.03) | (0.11) | (0.03) |
| 0.34 | 0.17 | 0.38 | 0.33 | 0.14 | 0.43 |
| (0.10) | (0.14) | (0.06) | (0.05) | (0.15) | (0.04) |
| 0.26 | 0.06 | 0.34 | 0.29 | 0.01 | 0.40 |
| (0.09) | (0.13) | (0.05) | (0.04) | (0.14) | (0.04) |
| 0.33 | 0.19 | 0.36 | 0.43 | 0.42 | 0.53 |
| (0.12) | (0.07) | (0.09) | (0.10) | (0.21) | (0.15) |
| 0.26 | 0.15 | 0.31 | 0.37 | 0.30 | 0.45 |
| (0.11) | (0.06) | (0.08) | (0.09) | (0.19) | (0.14) |
| −0.32 | −0.45 | −0.42 | −0.39 | 0.00 | −0.23 |
| (0.13) | (0.07) | (0.09) | (0.10) | (0.21) | (0.16) |
| −0.15 | −0.36 | −0.30 | −0.26 | 0.29 | −0.02 |
| (0.11) | (0.06) | (0.08) | (0.09) | (0.19) | (0.14) |
| −0.13 | −0.10 | −0.13 | −0.08 | −0.15 | −0.22 |
| (0.07) | (0.03) | (0.05) | (0.05) | (0.11) | (0.08) |
| −0.15 | −0.12 | −0.15 | −0.11 | −0.20 | −0.25 |
| (0.06) | (0.03) | (0.04) | (0.05) | (0.10) | (0.07) |
| 0.12 | 0.36 | 0.19 | 0.05 | −0.27 | −0.09 |
| (0.07) | (0.04) | (0.05) | (0.06) | (0.12) | (0.09) |
| 0.05 | 0.32 | 0.14 | −0.01 | −0.39 | −0.18 |
| (0.06) | (0.03) | (0.04) | (0.05) | (0.10) | (0.07) |
| 1.93 | | | 1.94 | 0.74 | 5.19 |
| (0.14) | c | c | (0.13) | (0.05) | (0.40) |

**Table 5.3**
(continued)

| Elasticity[a] | Version[b] | Canada | France | Italy | Japan |
|---|---|---|---|---|---|
| | (b) | 1.40 | 2.31 | 1.59 | 2.22 |
| | | (0.09) | (0.15) | (0.11) | (0.15) |
| $\eta_{32}$ | (a) | −0.04 | −0.28 | −0.11 | −0.29 |
| | | (0.10) | (0.15) | (0.11) | (0.15) |
| | (b) | 0.08 | −0.35 | −0.15 | −0.35 |
| | | (0.09) | (0.13) | (0.10) | (0.14) |
| $\eta_{33}$ | (a) | −1.21 | −1.42 | −1.26 | −1.42 |
| | | (0.08) | (0.13) | (0.09) | (0.13) |
| | (b) | −1.19 | −1.38 | −1.23 | −1.38 |
| | | (0.08) | (0.12) | (0.09) | (0.12) |
| $\eta_{34}$ | (a) | −0.14 | −0.60 | −0.22 | −0.50 |
| | | (0.06) | (0.10) | (0.07) | (0.10) |
| | (b) | −0.13 | −0.58 | −0.20 | −0.48 |
| | | (0.06) | (0.10) | (0.07) | (0.10) |
| $\eta_{41}$ | (a) | 0.04 | 0.21 | 0.08 | 0.13 |
| | | (0.02) | (0.03) | (0.02) | (0.02) |
| | (b) | 0.02 | 0.19 | 0.06 | 0.11 |
| | | (0.02) | (0.02) | (0.02) | (0.02) |
| $\eta_{42}$ | (a) | 0.11 | 0.01 | 0.07 | 0.03 |
| | | (0.02) | (0.02) | (0.02) | (0.02) |
| | (b) | 0.09 | −0.02 | 0.05 | 0.02 |
| | | (0.02) | (0.02) | (0.02) | (0.01) |
| $\eta_{43}$ | (a) | −0.02 | −0.07 | −0.03 | −0.05 |
| | | (0.01) | (0.01) | (0.01) | (0.01) |
| | (b) | −0.02 | −0.08 | −0.03 | −0.05 |
| | | (0.01) | (0.01) | (0.01) | (0.01) |
| $\eta_{44}$ | (a) | −0.12 | −0.14 | −0.12 | −0.11 |
| | | (0.02) | (0.03) | (0.02) | (0.02) |
| | (b) | −0.08 | −0.10 | −0.08 | −0.07 |
| | | (0.02) | (0.02) | (0.02) | (0.02) |

| Netherlands | Norway | Sweden | U.K. | U.S. | West Germany |
|---|---|---|---|---|---|
| 1.93 | | | 1.94 | 0.74 | 5.12 |
| (0.13) | [c] | [c] | (0.12) | (0.05) | (0.36) |
| −0.26 | | | −0.20 | −0.07 | −0.99 |
| (0.13) | [c] | [c] | (0.12) | (0.05) | (0.36) |
| −0.32 | | | −0.25 | −0.09 | −1.14 |
| (0.12) | [c] | [c] | (0.11) | (0.04) | (0.33) |
| −1.36 | | | −1.33 | −0.97 | −2.16 |
| (0.12) | [c] | [c] | (0.11) | (0.04) | (0.32) |
| −1.33 | | | −1.30 | −0.96 | −2.06 |
| (0.11) | [c] | [c] | (0.10) | (0.04) | (0.30) |
| −0.30 | | | −0.41 | 0.30 | −2.04 |
| (0.09) | [c] | [c] | (0.08) | (0.03) | (0.24) |
| −0.28 | | | −0.39 | 0.30 | −2.00 |
| (0.09) | [c] | [c] | (0.08) | (0.03) | (0.24) |
| 0.07 | 0.02 | 0.14 | 0.18 | 0.02 | 0.22 |
| (0.02) | (0.02) | (0.02) | (0.03) | (0.02) | (0.02) |
| 0.05 | 0.01 | 0.13 | 0.16 | 0.00 | 0.20 |
| (0.02) | (0.02) | (0.02) | (0.02) | (0.02) | (0.02) |
| 0.03 | 0.15 | 0.07 | 0.02 | −0.03 | −0.02 |
| (0.02) | (0.02) | (0.02) | (0.02) | (0.02) | (0.02) |
| 0.01 | 0.13 | 0.05 | 0.00 | −0.05 | −0.04 |
| (0.01) | (0.01) | (0.01) | (0.02) | (0.01) | (0.01) |
| −0.03 | −0.10 | −0.11 | −0.06 | 0.09 | −0.09 |
| (0.01) | (0.01) | (0.01) | (0.01) | (0.01) | (0.01) |
| −0.03 | −0.10 | −0.11 | −0.06 | 0.09 | −0.09 |
| (0.01) | (0.01) | (0.01) | (0.01) | (0.01) | (0.01) |
| −0.06 | −0.07 | −0.11 | −0.14 | −0.07 | −0.11 |
| (0.02) | (0.02) | (0.02) | (0.03) | (0.02) | (0.02) |
| −0.03 | −0.04 | −0.07 | −0.10 | −0.04 | −0.07 |
| (0.02) | (0.02) | (0.02) | (0.02) | (0.02) | (0.02) |

**Table 5.4**
Own-price elasticities (ten countries, dummy variables on second-order terms, 1959–1974)

| Elasticity | Canada | France | Italy | Japan | Netherlands |
|---|---|---|---|---|---|
| $\eta_{11}$ | −2.91 | −1.44 | −0.97 | −0.08 | 0.30 |
| | (0.45) | (0.19) | (0.27) | (0.98) | (0.48) |
| $\eta_{22}$ | −0.39 | −0.68 | −0.76 | 0.26 | 0.36 |
| | (0.16) | (0.28) | (0.18) | (0.25) | (0.36) |
| $\eta_{33}$ | −1.18 | −0.45 | −0.29 | 1.11 | −2.77 |
| | (0.35) | (0.23) | (0.44) | (1.53) | (0.19) |
| $\eta_{44}$[a] | −0.15 | −0.32 | −0.25 | −0.22 | −0.15 |

[a] Standard errors could not easily be calculated for this elasticity.
[b] Almost no natural gas is consumed in the industrial sectors of Norway and Sweden, so that these elasticities are meaningless.

| Norway | Sweden | U.K. | U.S. | West Germany |
|---|---|---|---|---|
| 1.68 | 0.03 | −0.93 | 0.41 | −2.25 |
| (0.66) | (0.56) | (0.21) | (1.58) | (0.23) |
| −0.09 | 0.77 | −0.06 | −0.30 | −0.90 |
| (0.13) | (0.32) | (0.24) | (0.31) | (0.44) |
| [b] | [b] | −1.34 | −0.27 | −3.89 |
| | | (0.16) | (0.52) | (1.64) |
| 0.11 | −0.04 | −0.33 | −0.14 | −0.23 |

**Table 5.5**
Own-price elasticities (Europe and Japan, country dummy variables, data at three-year intervals)

| Elasticity | France | Italy | Japan | Netherlands |
|---|---|---|---|---|
| $\eta_{11}$ | −0.96 | −1.34 | −1.20 | −1.48 |
| | (0.13) | (0.24) | (0.19) | (0.29) |
| $\eta_{22}$ | −0.05 | −0.17 | −0.04 | −0.08 |
| | (0.20) | (0.16) | (0.20) | (0.24) |
| $\eta_{33}$ | −1.67 | −1.43 | −1.67 | −1.59 |
| | (0.26) | (0.19) | (0.26) | (0.24) |
| $\eta_{44}$ | −0.08 | −0.07 | −0.06 | −0.02 |
| | (0.03) | (0.03) | (0.03) | (0.02) |

[a] Almost no natural gas is consumed in the industrial sectors of Norway and Sweden, so that these elasticities are meaningless.

| Norway | Sweden | U.K. | West Germany |
|---|---|---|---|
| −1.80 | −1.14 | −1.03 | −0.98 |
| (0.43) | (0.18) | (0.14) | (0.13) |
| −0.24 | −0.14 | −0.07 | 0.25 |
| (0.13) | (0.17) | (0.19) | (0.29) |
| [a] | [a] | −1.53 | −2.76 |
| | | (0.22) | (0.65) |
| −0.03 | −0.06 | −0.08 | −0.06 |
| (0.03) | (0.03) | (0.03) | (0.03) |

**Table 5.6**
Energy price indices derived from preferred fuel-choice model (U.S. = 1 in 1970)

| | Canada | France | Italy | Japan | Netherlands |
|---|---|---|---|---|---|
| 1955 | 1.0707 | 2.1349 | 3.6862 | 3.4706 | 7.5825 |
| 1956 | 1.2588 | 2.0541 | 3.5704 | 3.3103 | 6.9176 |
| 1957 | 1.2738 | 2.0432 | 3.5807 | 3.1635 | 6.9216 |
| 1958 | 1.2187 | 1.9325 | 3.4128 | 3.1860 | 6.9786 |
| 1959 | 1.1550 | 1.9805 | 3.3691 | 3.1010 | 6.6800 |
| 1960 | 1.1574 | 1.8590 | 3.3315 | 2.9548 | 6.4169 |
| 1961 | 1.1452 | 1.8086 | 3.0781 | 2.8187 | 6.1131 |
| 1962 | 1.1040 | 1.7484 | 2.9146 | 2.7179 | 5.4860 |
| 1963 | 1.1039 | 1.6921 | 2.6765 | 2.5977 | 5.0274 |
| 1964 | 1.0480 | 1.6991 | 2.4554 | 2.4869 | 4.5576 |
| 1965 | 1.0633 | 1.6738 | 2.3863 | 2.7431 | 4.3033 |
| 1966 | 0.9826 | 1.6495 | 2.3320 | 2.6305 | 4.3704 |
| 1967 | 0.9578 | 1.6738 | 2.2923 | 2.5159 | 4.0833 |
| 1968 | 0.9648 | 1.7420 | 2.2749 | 2.4010 | 3.9185 |
| 1969 | 0.9338 | 1.7254 | 2.1749 | 2.3594 | 3.5203 |
| 1970 | 0.9357 | 1.7286 | 2.0375 | 2.2061 | 3.2744 |
| 1971 | 0.9248 | 1.7568 | 1.9222 | 2.1358 | 3.0467 |
| 1972 | 0.8438 | 1.7432 | 1.8195 | 2.0930 | 2.7999 |
| 1973 | 0.8525 | 1.6517 | 1.7615 | 2.0201 | 2.4709 |
| 1974 | 1.0646 | 1.6831 | 1.6346 | na | 2.6157 |

| Norway | Sweden | U.K. | U.S. | West Germany |
|---|---|---|---|---|
| 1.1147 | 2.7962 | 2.1636 | na | 3.3351 |
| 1.1020 | 2.7825 | 2.1752 | 1.4599 | 3.1851 |
| 1.1235 | 2.8569 | 2.2623 | 1.4532 | 3.2232 |
| 1.0727 | 2.6824 | 2.2512 | na | 3.2704 |
| 1.0135 | 2.5866 | 2.2051 | 1.3698 | 3.1876 |
| 0.9468 | 2.6763 | 2.1629 | 1.3466 | 3.2221 |
| 0.9238 | 2.6258 | 2.2206 | 1.3276 | 3.0536 |
| 0.9515 | 2.6846 | 2.1628 | 1.2956 | 2.9061 |
| 0.9437 | 2.4719 | 2.1808 | 1.2372 | 2.8344 |
| 0.9217 | 2.3112 | 2.1125 | 1.1897 | 2.7768 |
| 0.8801 | 2.1428 | 2.0970 | 1.1579 | 2.8191 |
| 0.8990 | 2.0422 | 2.0547 | 1.1111 | 2.7484 |
| 0.9428 | 1.9952 | 2.0402 | 1.0887 | 2.6623 |
| 0.8938 | 1.8700 | 1.9372 | 1.0495 | 2.8022 |
| 0.9025 | 1.7619 | 1.8422 | 1.0088 | 2.5817 |
| 0.9264 | 1.6633 | 1.8091 | 1.0000 | 2.4469 |
| 0.9612 | 1.6143 | 1.9275 | 1.0603 | 2.3168 |
| 1.0002 | 1.5811 | 1.8696 | 1.0900 | 2.3266 |
| 0.9404 | 1.4522 | 1.7409 | 1.1197 | 2.2439 |
| 1.0988 | 1.6794 | 2.0924 | 1.4096 | 2.2444 |

**Table 5.7**
Parameter estimates of factor-share model

| parameters[a] | (1) Homothetic, ten countries, 1963-1973: regionally homogeneous | (2) Nonhomothetic, ten countries, 1963-1973: regionally homogeneous | (3) Nonhomothetic, ten countries, 1963-1973 |
|---|---|---|---|
| $\alpha_K$ | 0.3761 (0.0171) | 0.2137 (0.0399) | |
| $\alpha_K D_1$ | | | 0.3138 (0.1881) |
| $\alpha_K D_2$ | | | 0.2887 (0.2254) |
| $\alpha_K D_3$ | | | 0.2451 (0.2056) |
| $\alpha_K D_4$ | | | 0.3468 (0.2520) |
| $\alpha_K D_5$ | | | 0.2492 (0.1658) |
| $\alpha_K D_6$ | | | 0.3018 (0.1281) |
| $\alpha_K D_7$ | | | 0.2452 (0.1658) |
| $\alpha_K D_8$ | | | 0.0852 (0.2247) |
| $\alpha_K D_9$ | | | 0.2947 (0.2718) |
| $\alpha_K D_{10}$ | | | 0.3841 (0.2436) |
| $\alpha_L$ | 0.5754 (0.0160) | 0.7021 (0.0383) | |
| $\alpha_L D_1$ | | | 0.5749 (0.1780) |
| $\alpha_L D_2$ | | | 0.5950 (0.2137) |
| $\alpha_L D_3$ | | | 0.6232 (0.1950) |
| $\alpha_L D_4$ | | | 0.5139 (0.2389) |
| $\alpha_L D_5$ | | | 0.6438 (0.1577) |
| $\alpha_L D_6$ | | | 0.5961 (0.1213) |
| $\alpha_L D_7$ | | | 0.6358 (0.1571) |
| $\alpha_L D_8$ | | | 0.7867 (0.2131) |
| $\alpha_L D_9$ | | | 0.5842 (0.2572) |
| $\alpha_L D_{10}$ | | | 0.4978 (0.2310) |
| $\alpha_E$ | 0.0485 (0.0022) | 0.0841 (0.0040) | |
| $\alpha_E D_1$ | | | 0.1112 (0.0332) |
| $\alpha_E D_2$ | | | 0.1163 (0.0412) |
| $\alpha_E D_3$ | | | 0.1317 (0.0387) |
| $\alpha_E D_4$ | | | 0.1393 (0.0447) |
| $\alpha_E D_5$ | | | 0.1070 (0.0316) |
| $\alpha_E D_6$ | | | 0.1021 (0.0223) |
| $\alpha_E D_7$ | | | 0.1190 (0.0283) |
| $\alpha_E D_8$ | | | 0.1280 (0.0412) |
| $\alpha_E D_9$ | | | 0.1211 (0.0480) |
| $\alpha_E D_{10}$ | | | 0.1182 (0.0424) |

[a] $K$ = capital, $L$ = labor, $E$ = energy.

| (4) Nonhomothetic, U.S. and Canada, 1963-1973 | (5) Nonhomothetic, Europe and Japan, 1963-1973 | (6) Nonhomothetic, ten countries, data at three-year intervals |
|---|---|---|
| 0.8375 (0.3930) | 0.0649 (0.2584) | 0.8517 (0.3252) |
| 1.0104 (0.5869) | 0.0432 (0.2356) | 0.9604 (0.3908) |
| | 0.0969 (0.2890) | 0.8540 (0.3576) |
| | 0.0919 (0.1899) | 1.1027 (0.4370) |
| | 0.1748 (0.1465) | 0.7229 (0.2867) |
| | 0.0829 (0.1898) | 0.6762 (0.2190) |
| | −0.1353 (0.2575) | 0.7191 (0.2850) |
| | 0.1438 (0.2792) | 0.7403 (0.3916) |
| | | 1.0511 (0.4689) |
| | | 1.1119 (0.4268) |
| 0.1920 (0.3614) | 0.7559 (0.2479) | 0.0624 (0.3090) |
| 0.0710 (0.5393) | 0.7677 (0.2261) | −0.0456 (0.3721) |
| | 0.6934 (0.2773) | 0.0429 (0.3407) |
| | 0.7542 (0.1828) | −0.2055 (0.4163) |
| | 0.6880 (0.1405) | 0.1982 (0.2745) |
| | 0.7521 (0.1820) | 0.2400 (0.2082) |
| | 0.9448 (0.2471) | 0.1839 (0.2713) |
| | 0.6701 (0.2679) | 0.1620 (0.3731) |
| | | −0.1368 (0.4455) |
| | | −0.1961 (0.4065) |
| −0.0295 (0.9794) | 0.1791 (0.0447) | 0.0859 (0.0608) |
| −0.0813 (0.9003) | 0.1891 (0.0400) | 0.0852 (0.0748) |
| | 0.2097 (0.0500) | 0.1031 (0.0707) |
| | 0.1539 (0.0361) | 0.1028 (0.0843) |
| | 0.1375 (0.0245) | 0.0790 (0.0600) |
| | 0.1650 (0.3332) | 0.0838 (0.0424) |
| | 0.1908 (0.0448) | 0.0969 (0.0548) |
| | 0.1862 (0.0480) | 0.0977 (0.0762) |
| | | 0.0857 (0.0883) |
| | | 0.0843 (0.0825) |

**Table 5.7**
(continued)

| parameters[a] | (1) Homothetic, ten countries, 1963–1973: regionally homogeneous | (2) Nonhomothetic, ten countries, 1963–1973: regionally homogeneous | (3) Nonhomothetic, ten countries, 1963–1973 |
|---|---|---|---|
| $\gamma_{KK}$ | 0.0685 (0.0211) | 0.0525 (0.0190) | 0.0467 (0.0244) |
| $\gamma_{KL}$ | −0.0697 (0.0193) | −0.0569 (0.0179) | −0.0411 (0.0228) |
| $\gamma_{KE}$ | 0.0011 (0.0036) | 0.0044 (0.0027) | −0.0056 (0.0030) |
| $\gamma_{LK}$ | −0.0697 (0.0193) | −0.0569 (0.0179) | −0.0411 (0.0228) |
| $\gamma_{LL}$ | 0.0775 (0.0180) | 0.0675 (0.0172) | 0.0422 (0.0215) |
| $\gamma_{LE}$ | −0.0078 (0.0024) | −0.0106 (0.0018) | −0.0011 (0.0036) |
| $\gamma_{EK}$ | 0.0011 (0.0036) | 0.0044 (0.0027) | −0.0056 (0.0030) |
| $\gamma_{EL}$ | −0.0078 (0.0024) | −0.0106 (0.0018) | −0.0011 (0.0036) |
| $\gamma_{EE}$ | 0.0067 (0.0021) | 0.0062 (0.0016) | 0.0067 (0.0028) |
| $\gamma_{QK}$ | | 0.0293 (0.0066) | 0.0216 (0.0333) |
| $\gamma_{QL}$ | | −0.0229 (0.0063) | −0.0104 (0.0315) |
| $\gamma_{QE}$ | | −0.0065 (0.0007) | −0.0113 (0.0059) |
| | **RSQ** | **RSQ** | **RSQ** |
| Equation 1 | 0.228 | 0.095 | 0.859 |
| Equation 2 | 0.214 | 0.126 | 0.864 |

| (4) Nonhomothetic, U.S. and Canada, 1963–1973 | (5) Nonhomothetic, Europe and Japan, 1963–1973 | (6) Nonhomothetic, ten countries, data at three-year intervals |
|---|---|---|
| −0.1068 (0.0795) | 0.0732 (0.0270) | −0.0387 (0.0434) |
| 0.0957 (0.0728) | −0.0628 (0.0254) | 0.0446 (0.0403) |
| 0.0111 (0.0083) | −0.0104 (0.0032) | −0.0059 (0.0054) |
| 0.0957 (0.0728) | −0.0628 (0.0254) | 0.0446 (0.0403) |
| −0.0808 (0.0673) | 0.0585 (0.0243) | −0.0406 (0.0382) |
| −0.0149 (0.0087) | 0.0043 (0.0039) | −0.0041 (0.0070) |
| 0.0111 (0.0083) | −0.0104 (0.0032) | −0.0059 (0.0054) |
| −0.0149 (0.0087) | 0.0043 (0.0039) | −0.0041 (0.0070) |
| 0.0038 (0.0042) | 0.0061 (0.0031) | 0.0100 (0.0047) |
| −0.0648 (0.0727) | 0.0539 (0.0380) | −0.0701 (0.0574) |
| 0.0513 (0.0667) | −0.0335 (0.0364) | 0.0771 (0.0545) |
| 0.0135 (0.0091) | −0.0203 (0.0065) | −0.0069 (0.0108) |
| **RSQ** | **RSQ** | **RSQ** |
| 0.259 | 0.880 | 0.871 |
| 0.165 | 0.885 | 0.877 |

**Table 5.8**
Elasticities of substitution for capital, labor, and energy

| Elasticity | Model[a] | Canada | France | Italy | Japan |
|---|---|---|---|---|---|
| $\sigma_{KK}$ | (a) | −0.99 | −0.91 | −1.14 | −0.68 |
| | | (0.12) | (0.11) | (0.14) | (0.08) |
| | (b) | −1.75 | −0.79 | −0.98 | −0.59 |
| | | (0.39) | (0.13) | (0.16) | (0.09) |
| $\sigma_{KL}$ | (a) | 0.82 | 0.82 | 0.81 | 0.81 |
| | | (0.10) | (0.10) | (0.11) | (0.11) |
| | (b) | 1.43 | 0.72 | 0.70 | 0.70 |
| | | (0.32) | (0.11) | (0.12) | (0.12) |
| $\sigma_{KE}$ | (a) | 0.76 | 0.76 | 0.82 | 0.86 |
| | | (0.13) | (0.13) | (0.09) | (0.07) |
| | (b) | 1.48 | 0.56 | 0.67 | 0.74 |
| | | (0.36) | (0.13) | (0.10) | (0.08) |
| $\sigma_{LL}$ | (a) | −0.84 | −0.90 | −0.79 | −1.32 |
| | | (0.09) | (0.09) | (0.09) | (0.15) |
| | (b) | −1.33 | −0.83 | −0.73 | −1.22 |
| | | (0.27) | (0.11) | (0.09) | (0.17) |
| $\sigma_{LE}$ | (a) | 0.96 | 0.95 | 0.97 | 0.96 |
| | | (0.14) | (0.15) | (0.09) | (0.13) |
| | (b) | 0.42 | 1.17 | 1.11 | 1.15 |
| | | (0.35) | (0.16) | (0.10) | (0.14) |
| $\sigma_{EE}$ | (a) | −15.86 | −16.20 | −10.96 | −11.17 |
| | | (1.07) | (1.12) | (0.50) | (0.51) |
| | (b) | −16.96 | −16.45 | −11.06 | −11.28 |
| | | (1.58) | (1.22) | (0.54) | (0.56) |

[a] Model version (a) ten countries pooled, (b) U.S. and Canada pooled separately from Europe and Japan.

| Netherlands | Norway | Sweden | U.K. | U.S. | West Germany |
|---|---|---|---|---|---|
| −1.35 | −1.12 | −1.39 | −2.18 | −0.95 | −0.57 |
| (0.18) | (0.14) | (0.18) | (0.37) | (0.11) | (0.07) |
| −1.16 | −0.97 | −1.20 | −1.78 | −1.66 | −0.49 |
| (0.19) | (0.15) | (0.20) | (0.41) | (0.37) | (0.08) |
| 0.81 | 0.82 | 0.80 | 0.77 | 0.82 | 0.81 |
| (0.11) | (0.10) | (0.11) | (0.13) | (0.10) | (0.11) |
| 0.70 | 0.71 | 0.69 | 0.64 | 1.41 | 0.71 |
| (0.12) | (0.12) | (0.12) | (0.15) | (0.31) | (0.12) |
| 0.77 | 0.78 | 0.80 | 0.66 | 0.61 | 0.82 |
| (0.12) | (0.12) | (0.11) | (0.18) | (0.21) | (0.10) |
| 0.59 | 0.59 | 0.63 | 0.36 | 1.77 | 0.66 |
| (0.13) | (0.13) | (0.11) | (0.19) | (0.58) | (0.10) |
| −0.65 | −0.76 | −0.65 | −0.38 | −0.81 | −1.42 |
| (0.08) | (0.08) | (0.07) | (0.05) | (0.08) | (0.16) |
| −0.60 | −0.70 | −0.60 | −0.34 | −1.29 | −1.30 |
| (0.08) | (0.09) | (0.08) | (0.05) | (0.26) | (0.18) |
| 0.97 | 0.96 | 0.96 | 0.97 | 0.93 | 1.94 |
| (0.09) | (0.11) | (0.08) | (0.08) | (0.23) | (0.19) |
| 1.11 | 1.14 | 1.10 | 1.10 | 0.05 | 1.23 |
| (0.10) | (0.12) | (0.09) | (0.09) | (0.56) | (0.21) |
| −12.25 | −13.75 | −10.77 | −13.10 | −24.21 | −15.82 |
| (0.61) | (0.78) | (0.47) | (0.70) | (2.93) | (1.06) |
| −12.39 | −13.92 | −10.88 | −13.25 | −27.21 | −16.05 |
| (0.67) | (0.86) | (0.52) | (0.78) | (4.35) | (1.16) |

**Table 5.9**
Price elasticities of demand for capital, labor, and energy

| Elasticity | Model[a] | Canada | France | Italy | Japan |
|---|---|---|---|---|---|
| $\eta_{KK}$ | (a) | −0.45 | −0.43 | −0.47 | −0.37 |
| | | (0.05) | (0.05) | (0.06) | (0.05) |
| | (b) | −0.78 | −0.37 | −0.41 | −0.32 |
| | | (0.18) | (0.06) | (0.06) | (0.05) |
| $\eta_{KL}$ | (a) | 0.41 | 0.39 | 0.41 | 0.31 |
| | | (0.05) | (0.05) | (0.05) | (0.04) |
| | (b) | 0.71 | 0.35 | 0.36 | 0.27 |
| | | (0.16) | (0.05) | (0.06) | (0.05) |
| $\eta_{KE}$ | (a) | 0.04 | 0.04 | 0.06 | 0.06 |
| | | (0.01) | (0.01) | (0.01) | (0.01) |
| | (b) | 0.08 | 0.03 | 0.05 | 0.06 |
| | | (0.02) | (0.01) | (0.01) | (0.01) |
| $\eta_{LK}$ | (a) | 0.37 | 0.38 | 0.33 | 0.44 |
| | | (0.05) | (0.05) | (0.04) | (0.06) |
| | (b) | 0.64 | 0.34 | 0.29 | 0.38 |
| | | (0.15) | (0.05) | (0.05) | (0.07) |
| $\eta_{LL}$ | (a) | −0.42 | −0.43 | −0.41 | −0.51 |
| | | (0.04) | (0.04) | (0.04) | (0.06) |
| | (b) | −0.66 | −0.40 | −0.37 | −0.46 |
| | | (0.14) | (0.05) | (0.05) | (0.06) |
| $\eta_{LE}$ | (a) | 0.05 | 0.05 | 0.07 | 0.07 |
| | | (0.01) | (0.01) | (0.01) | (0.01) |
| | (b) | 0.02 | 0.06 | 0.08 | 0.09 |
| | | (0.02) | (0.01) | (0.01) | (0.01) |
| $\eta_{EK}$ | (a) | 0.34 | 0.36 | 0.34 | 0.47 |
| | | (0.06) | (0.06) | (0.04) | (0.04) |
| | (b) | 0.66 | 0.26 | 0.28 | 0.40 |
| | | (0.16) | (0.06) | (0.04) | (0.04) |
| $\eta_{EL}$ | (a) | 0.48 | 0.46 | 0.50 | 0.37 |
| | | (0.07) | (0.07) | (0.05) | (0.05) |
| | (b) | 0.21 | 0.56 | 0.57 | 0.44 |
| | | (0.17) | (0.08) | (0.05) | (0.05) |
| $\eta_{EE}$ | (a) | −0.82 | −0.82 | −0.84 | −0.84 |
| | | (0.05) | (0.06) | (0.04) | (0.04) |
| | (b) | −0.87 | −0.83 | −0.84 | −0.84 |
| | | (0.08) | (0.06) | (0.04) | (0.04) |

[a] Model version (a) ten countries pooled, (b) U.S. and Canada pooled separately from Europe and Japan.

| Netherlands | Norway | Sweden | U.K. | U.S. | West Germany |
|---|---|---|---|---|---|
| −0.50 | −0.47 | −0.51 | −0.56 | −0.44 | −0.33 |
| (0.07) | (0.06) | (0.07) | (0.10) | (0.05) | (0.04) |
| −0.43 | −0.41 | −0.43 | −0.46 | −0.71 | −0.29 |
| (0.07) | (0.06) | (0.07) | (0.10) | (0.17) | (0.05) |
| 0.45 | 0.42 | 0.44 | 0.52 | 0.42 | 0.30 |
| (0.06) | (0.05) | (0.06) | (0.09) | (0.05) | (0.04) |
| 0.39 | 0.37 | 0.38 | 0.44 | 0.71 | 0.26 |
| (0.07) | (0.06) | (0.06) | (0.10) | (0.16) | (0.04) |
| 0.05 | 0.05 | 0.06 | 0.04 | 0.02 | 0.04 |
| (0.01) | (0.01) | (0.01) | (0.01) | (0.01) | (0.01) |
| 0.04 | 0.04 | 0.05 | 0.02 | 0.06 | 0.03 |
| (0.01) | (0.01) | (0.01) | (0.01) | (0.02) | (0.01) |
| 0.30 | 0.34 | 0.29 | 0.20 | 0.38 | 0.47 |
| (0.04) | (0.04) | (0.04) | (0.03) | (0.04) | (0.06) |
| 0.26 | 0.30 | 0.26 | 0.17 | 0.65 | 0.42 |
| (0.05) | (0.05) | (0.05) | (0.04) | (0.14) | (0.07) |
| −0.36 | −0.40 | −0.36 | −0.26 | −0.41 | −0.52 |
| (0.04) | (0.04) | (0.04) | (0.03) | (0.04) | (0.06) |
| −0.34 | −0.37 | −0.33 | −0.23 | −0.65 | −0.47 |
| (0.04) | (0.05) | (0.04) | (0.04) | (0.13) | (0.07) |
| 0.06 | 0.06 | 0.08 | 0.06 | 0.03 | 0.05 |
| (0.01) | (0.01) | (0.01) | (0.01) | (0.01) | (0.01) |
| 0.08 | 0.07 | 0.09 | 0.07 | 0.00 | 0.06 |
| (0.01) | (0.01) | (0.01) | (0.01) | (0.02) | (0.01) |
| 0.29 | 0.33 | 0.29 | 0.17 | 0.28 | 0.48 |
| (0.04) | (0.05) | (0.04) | (0.05) | (0.10) | (0.06) |
| 0.22 | 0.25 | 0.23 | 0.09 | 0.82 | 0.38 |
| (0.05) | (0.05) | (0.04) | (0.05) | (0.27) | (0.06) |
| 0.54 | 0.50 | 0.55 | 0.66 | 0.47 | 0.34 |
| (0.05) | (0.06) | (0.05) | (0.06) | (0.12) | (0.07) |
| 0.62 | 0.59 | 0.62 | 0.75 | 0.03 | 0.45 |
| (0.06) | (0.06) | (0.05) | (0.06) | (0.29) | (0.08) |
| −0.84 | −0.83 | −0.84 | −0.84 | −0.75 | −0.82 |
| (0.04) | (0.05) | (0.04) | (0.04) | (0.09) | (0.05) |
| −0.84 | −0.84 | −0.84 | −0.84 | −0.85 | −0.85 |
| (0.05) | (0.05) | (0.04) | (0.05) | (0.14) | (0.06) |

**Table 5.10**
Index of scale economies and output elasticity of energy demand (all ten countries pooled)[a]

| Country | SCE | $\eta_{EQ}$ |
|---|---|---|
| Canada | 0.0015 (0.0080) | 0.785 (0.108) |
| France | 0.0086 (0.0383) | 0.783 (0.113) |
| Italy | 0.0032 (0.0335) | 0.855 (0.078) |
| Japan | 0.0105 (0.0487) | 0.849 (0.087) |
| Netherlands | −0.0124 (0.0170) | 0.818 (0.093) |
| Norway | 0.0056 (0.0172) | 0.807 (0.097) |
| Sweden | 0.0019 (0.0205) | 0.864 (0.070) |
| U.K. | 0.0041 (0.0327) | 0.778 (0.113) |
| U.S. | 0.0003 (0.0037) | 0.624 (0.188) |
| West Germany | 0.0012 (0.0316) | 0.761 (0.122) |

[a] Standard errors are computed based on constancy of shares and prices at their mean values. The standard error of SCE is thus computed from

$$\text{Var (SCE)} = \sum_{i=K}^{L,M} (\log P_i)^2 \text{ Var } (\gamma_{Qi}) + \sum_{i \neq j} \log P_i \log P_j \text{ Covar } (\gamma_{Qi}\gamma_{Qj}),$$

and the standard error of $\eta_{EQ}$ is computed from

$$\begin{aligned}\text{Var } (\eta_{EQ}) = {} & \text{Var (SCE)} + (2 \log P_E/S_E + 1/S_E^2) \text{ Var } (\gamma_{EQ}) \\ & + (2/S_E) \log P_K \text{ Covar } (\gamma_{QE}\gamma_{QK}) + (2/S_E) \log P_L \text{ Covar } (\gamma_{QE}\gamma_{QL}).\end{aligned}$$

**Table 5.11**
Total fuel-price elasticities (using preferred fuel-share and factor-share models)

| Elasticity[a] | Canada | France | Italy | Japan | Netherlands |
|---|---|---|---|---|---|
| $\eta_{11}$ | −1.89 | −1.29 | −1.63 | −1.49 | −1.78 |
| $\eta_{12}$ | 0.69 | 0.06 | 0.09 | 0.07 | 0.09 |
| $\eta_{13}$ | 1.08 | 0.40 | 0.76 | 0.60 | 0.92 |
| $\eta_{14}$ | −0.75 | 0.0 | −0.06 | −0.03 | −0.08 |
| $\eta_{21}$ | 0.31 | 0.11 | 0.07 | 0.09 | 0.09 |
| $\eta_{22}$ | −1.03 | −0.34 | −0.46 | −0.35 | −0.22 |
| $\eta_{23}$ | −0.29 | −0.13 | −0.10 | −0.13 | −0.16 |
| $\eta_{24}$ | 0.14 | −0.46 | −0.35 | −0.46 | −0.54 |
| $\eta_{31}$ | 1.25 | 1.96 | 1.39 | 1.95 | 1.73 |
| $\eta_{32}$ | −0.75 | −0.36 | −0.24 | −0.36 | −0.33 |
| $\eta_{33}$ | −0.41 | −1.54 | −1.37 | −1.54 | −1.48 |
| $\eta_{34}$ | −0.96 | −0.89 | −0.61 | −0.89 | −0.77 |
| $\eta_{41}$ | −0.15 | 0.0 | −0.01 | −0.01 | −0.02 |
| $\eta_{42}$ | 0.06 | −0.17 | −0.14 | −0.13 | −0.11 |
| $\eta_{43}$ | −0.17 | −0.12 | −0.10 | −0.09 | −0.08 |
| $\eta_{44}$ | −0.61 | −0.54 | −0.59 | −0.60 | −0.63 |

a. 1 = solid, 2 = liquid, 3 = gas, 4 = electricity.
b. Almost no natural gas is consumed in the industrial sectors of Norway and Sweden, so that these elasticities are meaningless.

| Norway | Sweden | U.K. | U.S. | West Germany |
|---|---|---|---|---|
| −2.15 | −1.44 | −1.35 | −2.24 | −1.31 |
| 0.15 | 0.07 | 0.06 | 0.92 | 0.05 |
| 1.33 | 0.54 | 0.45 | 1.50 | 0.41 |
| −0.16 | −0.02 | −0.02 | −1.03 | 0.01 |
| 0.05 | 0.08 | 0.10 | 0.90 | 0.13 |
| −0.56 | −0.44 | −0.37 | −1.17 | −0.06 |
| −0.09 | −0.12 | −0.12 | −0.88 | −0.20 |
| −0.24 | −0.37 | −0.44 | 0.30 | −0.70 |
| b | b | 1.64 | 0.65 | 4.73 |
| b | b | −0.30 | −0.38 | −0.93 |
| b | b | −1.44 | −0.67 | −2.34 |
| b | b | −0.74 | −0.43 | −2.29 |
| −0.02 | −0.01 | −0.01 | −0.13 | 0.01 |
| −0.10 | −0.13 | −0.16 | 0.04 | −0.14 |
| −0.09 | −0.10 | −0.11 | −0.13 | −0.10 |
| −0.62 | −0.60 | −0.56 | −0.63 | −0.59 |

**Table 5.12**
Elasticity of average cost of output with respect to price of energy and fuels

| Elasticity[a] | | Canada | France | Italy | Japan |
|---|---|---|---|---|---|
| A. U.S. and Canada pooled separately from Europe and Japan | | | | | |
| $\eta_{CE}$ | 1963 | 0.045 | 0.053 | 0.076 | 0.073 |
| | 1972 | 0.050 | 0.046 | 0.067 | 0.063 |
| $\eta_{C1}$ | 1963 | 0.006 | 0.017 | 0.009 | 0.012 |
| | 1972 | 0.004 | 0.008 | 0.010 | 0.007 |
| $\eta_{C2}$ | 1963 | 0.011 | 0.008 | 0.018 | 0.012 |
| | 1972 | 0.013 | 0.012 | 0.014 | 0.014 |
| $\eta_{C3}$ | 1963 | 0.004 | 0.005 | 0.006 | 0.006 |
| | 1972 | 0.007 | 0.004 | 0.007 | 0.002 |
| $\eta_{C4}$ | 1963 | 0.028 | 0.023 | 0.043 | 0.044 |
| | 1972 | 0.026 | 0.023 | 0.037 | 0.040 |
| B. All ten countries pooled | | | | | |
| $\eta_{CE}$ | 1963 | 0.053 | 0.053 | 0.076 | 0.072 |
| | 1972 | 0.047 | 0.048 | 0.068 | 0.063 |
| $\eta_{C1}$ | 1963 | 0.007 | 0.017 | 0.009 | 0.012 |
| | 1972 | 0.004 | 0.008 | 0.010 | 0.007 |
| $\eta_{C2}$ | 1963 | 0.011 | 0.008 | 0.018 | 0.012 |
| | 1972 | 0.012 | 0.012 | 0.014 | 0.014 |
| $\eta_{C3}$ | 1963 | 0.005 | 0.005 | 0.006 | 0.006 |
| | 1972 | 0.006 | 0.004 | 0.007 | 0.002 |
| $\eta_{C4}$ | 1963 | 0.031 | 0.023 | 0.043 | 0.042 |
| | 1972 | 0.024 | 0.024 | 0.037 | 0.040 |

[a] $E$ = energy; 1 = solid fuel, 2 = liquid fuel, 3 = gas, 4 = electricity.

| Netherlands | Norway | Sweden | U.K. | U.S. | West Germany |
|---|---|---|---|---|---|
| | | | | | |
| 0.069 | 0.062 | 0.078 | 0.065 | 0.029 | 0.051 |
| 0.060 | 0.062 | 0.066 | 0.059 | 0.032 | 0.043 |
| 0.010 | 0.004 | 0.015 | 0.019 | 0.003 | 0.014 |
| 0.004 | 0.006 | 0.010 | 0.008 | 0.002 | 0.005 |
| 0.013 | 0.016 | 0.012 | 0.010 | 0.002 | 0.007 |
| 0.004 | 0.016 | 0.015 | 0.016 | 0.003 | 0.007 |
| 0.00 | 0.00 | 0.001 | 0.005 | 0.005 | 0.001 |
| 0.011 | 0.00 | 0.00 | 0.005 | 0.006 | 0.003 |
| 0.045 | 0.044 | 0.050 | 0.030 | 0.019 | 0.030 |
| 0.041 | 0.040 | 0.041 | 0.030 | 0.020 | 0.028 |
| | | | | | |
| 0.069 | 0.064 | 0.075 | 0.065 | 0.033 | 0.051 |
| 0.060 | 0.062 | 0.066 | 0.060 | 0.028 | 0.044 |
| 0.009 | 0.004 | 0.014 | 0.019 | 0.003 | 0.014 |
| 0.004 | 0.006 | 0.010 | 0.008 | 0.002 | 0.005 |
| 0.013 | 0.016 | 0.012 | 0.010 | 0.002 | 0.007 |
| 0.004 | 0.016 | 0.015 | 0.017 | 0.003 | 0.007 |
| 0.000 | 0.00 | 0.001 | 0.005 | 0.006 | 0.001 |
| 0.011 | 0.00 | 0.00 | 0.005 | 0.006 | 0.003 |
| 0.046 | 0.044 | 0.048 | 0.030 | 0.022 | 0.030 |
| 0.041 | 0.039 | 0.041 | 0.030 | 0.018 | 0.028 |

**Table 5.13**
Summary of Elasticity Estimates

| Elasticity | Estimate | Table reference |
| --- | --- | --- |
| Factor inputs: elasticities of substitution | $\sigma_{KL} = 0.77$ to $0.82$<br>$\sigma_{KE} = 0.61$ to $0.86$<br>$\sigma_{LE} = 0.93$ to $0.97$ | 5.8 |
| Factor inputs: price elasticities | $\eta_{KK} = -0.33$ to $-0.56$<br>$\eta_{LL} = -0.26$ to $-0.52$<br>$\eta_{EE} = -0.75$ to $-0.84$<br>$\eta_{KE} = 0.02$ to $0.06$<br>$\eta_{LE} = 0.03$ to $0.08$ | 5.9 |
| Energy demand: output elasticity | $\eta_{EQ} = 0.62$ to $0.86$ | 5.10 |
| Fuels: own-price elasticities, partial | electricity: −0.07 to −0.16<br>oil: −0.81 to −1.10 for U.S. and Canada<br>−0.11 to −0.34 for Europe and Japan<br>gas: −0.33 to −0.52 for U.S. and Canada<br>−1.30 to −2.31 for Europe and Japan<br>coal: −1.80 to −2.17 for U.S. and Canada<br>−1.04 to −2.08 for Europe and Japan | 5.2 |
| Fuels: own-price elasticities, total | electricity: −0.54 to −0.63<br>oil: −1.03 to −1.117 for U.S. and Canada<br>−0.06 to −0.56 for Europe and Japan<br>gas: −0.41 to −0.67 for U.S. and Canada<br>−1.37 to −2.34 for Europe and Japan<br>coal: −1.89 to −2.24 for U.S. and Canada<br>−1.29 to −2.15 for Europe and Japan | 5.11 |

**Table 5.14**
Alternative estimates of industrial demand elasticities

| Elasticity | Country | Estimate | Source |
|---|---|---|---|
| Factor inputs: elasticities of substitution | U.S. | $\sigma_{KL} = 1.01$ | a |
| | | $\sigma_{KE} = -3.25$ | |
| | | $\sigma_{LE} = 0.64$ | |
| | U.S. (2-digit industries) | $\sigma_{KE} = -1.03$ to 2.02 | b |
| | | $\sigma_{LE} = 0.48$ to 2.88 (production workers) | |
| | | $\sigma_{LE} = -2.02$ to 5.59 (nonproduction workers) | |
| | Canada | $\sigma_{KL} = 0.72$ | c |
| | | $\sigma_{KE} = 0.42$ | |
| | | $\sigma_{LE} = 1.70$ | |
| | Canada | $\sigma_{KL} = 5.46$ | d |
| | | $\sigma_{KE} = -11.91$ | |
| | | $\sigma_{LE} = 4.89$ | |
| | Netherlands | $\sigma_{KL} = 1.09$ | e |
| | | $\sigma_{KE} = -4.41$ | |
| | | $\sigma_{LE} = 2.30$ | |
| | 9 industrialized countries | $\sigma_{KL} = 0.06$ to 0.52 | f |
| | | $\sigma_{KE} = 1.02$ to 1.07 | |
| | | $\sigma_{LE} = 0.72$ to 0.87 | |
| Factor Inputs: price elasticities | U.S. | $\eta_{KK} = -0.44$ | a |
| | | $\eta_{LL} = -0.45$ | |
| | | $\eta_{EE} = -0.49$ | |
| | | $\eta_{KE} = -0.15$ | |
| | | $\eta_{LE} = 0.03$ | |
| | U.S. (2-digit industries) | $\eta_{KK} = -0.67$ to $-1.16$ | b |
| | | $\eta_{LL} = -0.28$ to $-1.55$ | |
| | | $\eta_{EE} = -0.66$ to $-2.56$ | |
| | Canada | $\eta_{KK} = -0.79$ | c |
| | | $\eta_{LL} = -0.45$ | |
| | | $\eta_{EE} = -0.36$ | |
| | Canada | $\eta_{KK} = -0.31$ | d |
| | | $\eta_{LL} = -0.77$ | |
| | | $\eta_{EE} = -0.59$ | |

[a] Berndt and Wood (22).
[b] Halvorsen and Ford (79).
[c] Fuss and Waverman (62), translog.
[d] Fuss and Waverman (62), generalized Leontief.
[e] Magnus (114).
[f] Griffin and Gregory (71).
[g] Fuss (60).
[h] Nordhaus (123).
[i] Halvorsen (77).
[j] Mount, Chapman, and Tyrrell (121).
[k] Halvorsen (76).
[l] Griffin (69).
[m] Berndt and Watkins (21).

**Table 5.14**
(continued)

| Elasticity | Country | Estimate | Source |
|---|---|---|---|
| | Canada | $\eta_{KK} = -0.76$ | g |
| | | $\eta_{LL} = -0.49$ | |
| | | $\eta_{EE} = -0.49$ | |
| | | $\eta_{KE} = -0.05$ | |
| | | $\eta_{LE} = 0.55$ | |
| | Netherlands | $\eta_{KK} = -0.42$ | e |
| | | $\eta_{LL} = -0.46$ | |
| | | $\eta_{EE} = -0.29$ | |
| | 9 industrialized countries | $\eta_{KK} = -0.18$ to $-0.38$ | f |
| | | $\eta_{LL} = -0.12$ to $-0.27$ | |
| | | $\eta_{EE} = -0.79$ to $-0.80$ | |
| | | $\eta_{KE} \approx 0.13$ | |
| | | $\eta_{LE} \approx 0.11$ | |
| | 6 country composite | $\eta_{EE} = -0.30$ | h |
| Fuels: own-price elasticities, partial | U.S. | electricity: −0.66 | i |
| | | oil: −2.75 | |
| | | gas: −1.30 | |
| | | coal: −1.46 | |
| | U.S. | electricity: −0.14 (short run) | j |
| | | electricity: −1.20 (long run) | |
| | U.S. | electricity: −0.06 (short run) | k |
| | | electricity: −0.52 (long run) | |
| | 18 OECD countries pooled | electricity: −0.46 | l |
| | | oil: −0.71 | |
| | | gas: −1.14 | |
| | | coal: −1.17 | |
| Fuels: own-price elasticities, total | U.S. | electricity: −0.92 | i |
| | | oil: −2.82 | |
| | | gas: −1.47 | |
| | | coal: −1.52 | |
| | Canada | electricity: −0.74 | g |
| | | oil: −1.30 | |
| | | gas: −1.30 | |
| | | coal: −0.48 | |
| | Canada | gas: −0.60 | m |

# 6 The Transportation Demand for Energy

The dependence of energy demand on an energy-consuming stock of capital is more fundamental to the transportation sector than to the residential or industrial sectors. Residential energy demand, for example, will depend in the short run on the stock of fuel-burning appliances, as well as on a larger variety of appliances, such as heating units of different types, refrigerators, and air conditioners, and on such factors as the size and characteristics of the housing stock, the weather, and so forth. Because detailed data on the various types and characteristics of appliances and houses is not available, and because we have been more concerned with the long-run structure of demand, we treated energy in the residential and industrial sectors as a directly consumed good.

In the case of motor gasoline, however, demand for the fuel depends more explicitly and directly on a much more narrowly defined stock, namely, the stock of cars. In addition, detailed data (both price and quantity) for the stock and for its characteristics are available for a number of countries. For this reason we are able to take a very different approach in modeling the demand for motor gasoline, and treat it as a derived demand based on a set of estimated equations for the stock, use, and efficiency of automobiles. This approach has the advantage of providing dynamic elasticity estimates, that is, we can determine the response of gasoline demand over time to a change in price or income.

Our model of gasoline demand explains the annual consumption of gasoline as the ratio of the total traffic volume (which in turn is the product of average traffic volume per car and the stock of cars) to the average fuel efficiency of the stock of cars. Three equations determine the stock of cars, one for new registrations, a second for the depreciation rate, and the third an accounting identity. Two more equations complete the model; one explains average traffic volume per car, and the second explains average fuel efficiency.

The demands for other fuels used in the transportation sector should also depend on energy-consuming stocks. Diesel fuel demand, for example, should depend on the sizes and characteristics of the stocks of trucks and other commercial vehicles in use at any particular time. Because of a limited availability of data, however, we model this demand, as well as the demands for aviation gasoline and jet fuel, by using simple log-linear equations to relate them directly to prices and per capita GDP.

We have estimated the model of motor gasoline demand by pooling data for eleven countries: Belgium, Canada, France, Italy, the Netherlands, Norway, Sweden, Switzerland, the U.K., the U.S., and West Germany. We also estimate demand elasticities for gasoline in nine additional countries, Austria, Australia, Denmark, Finland, Greece, Turkey, Spain, Brazil, and Mexico, but here log-linear equations are estimated for the demands for diesel fuel, aviation gasoline, and jet fuel in all twenty countries.[1] We present the statistical results here, beginning with the model of gasoline demand.[2]

## 6.1 The Demand for Motor Gasoline

Before discussing the estimated equations of the model, it is useful to examine the behavior of gasoline prices and consumption over time and across countries. The per capita consumption of gasoline over time is shown for six countries in figures 6.1 and 6.2, and the

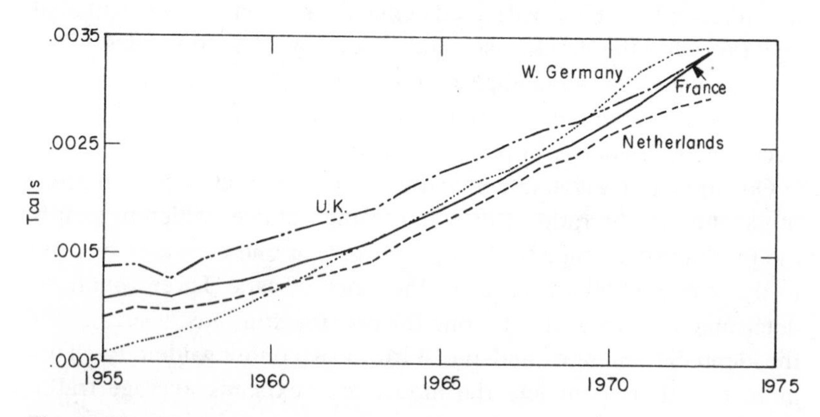


**Figure 6.1**
Per capita gasoline consumption of four European countries

1. Estimated demand equations for Greece, Turkey, Spain, Brazil, and Mexico are presented in the next chapter where we discuss the characteristics of energy demand in the developing countries.
2. All of the empirical results presented in this chapter were obtained by Ross Heide as part of his MIT Sloan School master's thesis. The results are described in more detail in Heide (82, 83).

average price is shown for the same countries in figure 6.3. Observe first that until 1973 the real price of gasoline has been falling steadily, while per capita consumption has been growing at an average rate of about 7 or 8 percent per year. This is not to say, of course, that the growth of consumption is due to falling prices; real per capita incomes have also been rising over this period. We can, however, observe a strong negative correlation between prices and consump-

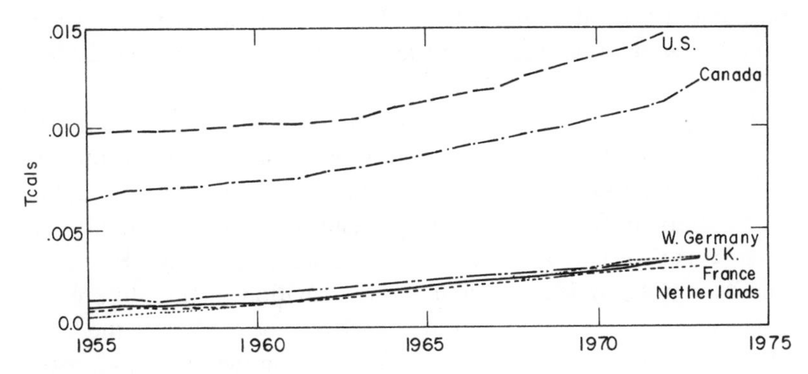


**Figure 6.2**
Per capita gasoline consumption of U.S. and Canada with the European countries

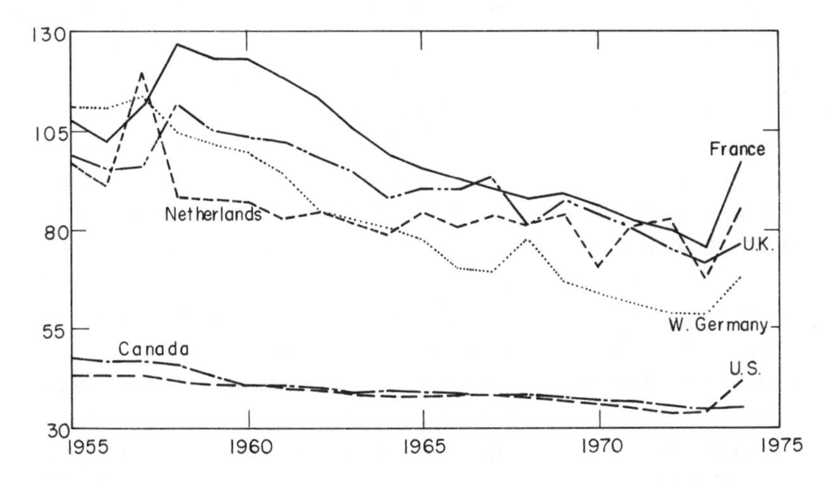


**Figure 6.3**
Average price of regular gasoline (1970 U.S. cents per gallon)

tion levels, both in terms of average values across countries and in terms of relative rates of change over time. For the European countries shown in those figures, consumption on average has been the lowest in the Netherlands and France, where prices on average have been the highest, while West Germany has experienced the most rapid growth rate of consumption combined with the greatest decline in price. Consumption in the U.S. and Canada has been four or five times as high as in the European countries, with prices only about a third as high. (In 1974, for example, per capita consumption in the European countries was between 60 and 100 gallons, while in the U.S. it was 420 gallons.)

The reason often cited for these intercountry differences in gasoline consumption (particularly between North America and Europe) is that differences in the distances between cities together with cultural differences or different tastes or habits lead to different preferences for larger cars or greater amounts of driving. However, this does not provide a meaningful explanation for the differences that we observe in consumption nor a basis for predicting the kinds of changes in consumption that are probable in the future as a result of changing prices and incomes. Tastes and habits are themselves likely to be functions of price, and the question is, To what extent and how rapidly will the average fuel efficiency and use of automobiles change in response to changes in the price of gasoline?

We try to answer this question with the model of gasoline demand developed in chapter 2. Because the speeds of adjustment of the different endogenous variables are an important characteristic of that model, estimation of the model requires variation in all of the variables across time as well as across countries. As can be seen in part from figures 6.1, 6.2, and 6.3, that variation is indeed present in the data. It should therefore be possible to elicit both long-run elasticities and their speeds of adjustment.

The model is estimated by pooling together all eleven countries in our sample. Individual equations were also estimated using alternative poolings of subgroupings of these countries. But pooling all of the countries gave the best results, and it was found that regional heterogeneity could be best treated through the use of regional dummy variables for the intercept parameters.

The final versions of the estimated equations used in the model are as follows (intercept parameters are shown in table 6.1):

New registrations:

$$NR/POP = \Sigma a_i D_i - \underset{(-2.39)}{79.9}\ PC - \underset{(-2.85)}{0.299}\ PG + \underset{(1.56)}{1.05}\,(GDP/POP) + \underset{(8.95)}{0.589}\,(NR_{t-1}/POP_{t-1}). \tag{6.1}$$

Depreciation rate:

$$R = \Sigma b_i D_i - \underset{(-2.23)}{2.93} \times 10^{-4} PC. \tag{6.2}$$

Traffic volume per car:

$$\log TVPC = \Sigma c_i D_i + \underset{(1.97)}{0.060} \log (GDP/POP) + \underset{(21.40)}{0.909} \log TVPC_{t-1}. \tag{6.3}$$

Average fuel efficiency:

$$\log (1/EFF) = \Sigma d_i D_i - \underset{(-2.30)}{0.110} \log PG + 0.923 \log (1/EFF_{t-1}), \tag{6.4}$$

$$\log (1/EFF - 0.489) = \Sigma d_i D_i - \underset{(-2.11)}{0.212} \log PG + 0.953 \log (1/EFF_{t-1} - 0.489). \tag{6.4a}$$

These equations explain new registrations, the depreciation rate, traffic volume per car, and average fuel efficiency, and together with the following identities for the stock of cars ($STK$) and consumption of gasoline ($Q$) complete the model:

$$STK_t = (1 - R)STK_{t-1} + NR_t, \tag{6.5}$$

$$Q_t = STK_t \cdot TVPC_t/EFF_t. \tag{6.6}$$

Recall from chapter 2 that equation (6.1) for new registrations was derived as a stock adjustment equation, with the coefficient of the stock term equal to the annual speed of stock adjustment net of the depreciation rate. In all versions of this equation, however, the stock term is either statistically insignificant or else positive.[3] This could mean that the speed of adjustment is approximately equal to or smaller than the depreciation rate (which seems unlikely), or simply that the dynamics of new registrations cannot in fact be explained by a stock adjustment hypothesis (so that the positive correlation between the two variables is an artifact resulting from positive trends in both). The stock term is therefore omitted from the final equation. Note that the price of cars and the price of gasoline both have negative coefficients and are statistically significant, while per capita gross domestic product has a positive coefficient but is not significant at the 5 percent level.

We explained in chapter 2 that in theory the depreciation rate could depend (positively) on per capita GDP as well as the price of cars. GDP, however, was not found to be significant, so the only explanatory variable in equation (6.2) is the price of cars. This is consistent with the fact that on the average depreciation rates differ considerably across countries, but they do not vary much over time. Note that all of the country intercept coefficients in equation (6.2) are highly significant and provide most of the explanation of the depreciation rate.

Traffic volume per car should depend positively on per capita GDP and negatively on the price of gasoline. We indeed find a positive dependence on GDP but no significant dependence on the price of gasoline, so the latter variable is dropped from the equation.[4]

Finally, two versions of the equation for average fuel efficiency are estimated. The first, equation (6.4), imposes no upper limit on the maximum achievable average fuel efficiency of the stock of cars, while the second equation (6.4a) imposes an upper limit of 40

3. When the term $(STK_{t-1}/POP_{t-1})$ is added to equation (6.1), its coefficient is positive but statistically insignificant. When this term is added but the lagged dependent variable is dropped, the estimated coefficient is significant and equal to 0.055. Since the average value of the depreciation rate is about 0.06, this would imply a speed of adjustment of 0.005, which is unrealistically small.

4. When $\log PG$ is added to equation (6.3), it appears with a positive coefficient (0.101) that is not significant at the 5 percent level.

miles per gallon ($2.04 \times 10^6$ km/Tcal) on average efficiency.[5] In both versions of the equation we find a strong negative dependence on the price of gasoline as expected, but one that operates with a long lag. (The median lag in equation (6.4) is 8.6 years, and in equation (6.4a) is 14.4 years.)

In fact we would expect there to be some upper limit on average fuel efficiency (at least given existing technologies), but unfortunately the range of the data does not permit us to estimate its value. This is a disadvantage of this version of the equation, since choosing the wrong value of the upper limit could amount to a misspecification that biases the other coefficient estimates. We therefore prefer equation (6.4), and use equation (6.4a) only if the range of data for the explanatory variables (both historical and forecasted) is such as to bring predicted average efficiency close to the exogenously imposed limit. This turns out to be the case only for Italy, and therefore, in simulating the model, equation (6.4a) is used for that country but (6.4) is used for the other ten countries.

Short- and long-run elasticities with respect to the price of gasoline, the price of cars, and per capita GDP have been calculated for the four endogenous variables explained by equations (6.1) through (6.4a), and these are shown in table 6.2. Observe that, according to the model, increases in the price of gasoline reduce consumption by reducing new registrations, which limits the stock of cars, and by increasing the average efficiency of the stock. Both of these effects, however, occur with long lags, so that we could expect to observe a large difference between short- and long-run elasticities of gasoline demand. The effect of an increase in per capita GDP also occurs in two ways, first by increasing new registrations, and second by increasing traffic volume per car.

Elasticities of gasoline demand with respect to price and per capita GDP are calculated by simulating the complete model, first using a set of base-case values for all of the exogenous variables, and then increasing one of the exogenous variables a fixed amount above its base-case trajectory. (Because the model is nonlinear, these demand elasticities will depend on all of the base-case values

5. Note that efficiency is measured in units of $10^6$ km/Tcal. To convert to miles per gallon, multiply by 20.

as well as the size of the deviation away from the base case.) Here the base-case simulation consists of a forecast in which the population of each country grows at its average historical rate, real GDP grows at 4 percent per year from the 1976 levels in all countries, the price of gasoline grows at 2 percent per year in real terms, and the price of automobiles is held constant in real terms. Elasticities are then computed by allowing either the price of gasoline or per capita GDP to rise 10 percent above its base-case path. These elasticities are shown for each country in tables 6.3 and 6.4.

Observe that according to this model the price elasticity of gasoline demand is very large in the long run—above 1 in every country, and close to 2 in Norway. (Elasticities will differ across countries because of differences in the country intercept coefficients in each equation and because of differences in the sample values of all of the endogenous and exogenous variables.) For the eleven-country aggregate the price elasticity reaches 0.5 after five years, exceeds 1 after fifteen years, and after twenty-five years reaches a value of 1.31. Thus, although the lags are considerable, we find gasoline demand to be much more price elastic in the long run than would be indicated by most earlier studies or by the conventional wisdom. As for the income elasticity of gasoline demand, the long-run values shown in table 6.4 are close to most earlier estimates, but what is surprising is the amount of time it would appear to take to reach the long-run value. For the eleven-country aggregate, this elasticity exceeds 0.5 only after nine or ten years. Thus, while steady 4 percent per year growth in GDP would lead to 3 or 3.5 percent per year growth in gasoline demand, short-run fluctuations in GDP would have little impact on demand.

It is interesting to compare the elasticities that have been derived from this model with those estimated by others. Houthakker, Verleger, and Sheehan (93) and Ramsey, Rasche, and Allen (144), working with time series data for the U.S., estimated long-run own-price elasticities of gasoline demand of −0.24 and −0.70, respectively, while Fuss and Waverman (62), using data for Canada, obtained estimates in the range −0.22 to −0.45. It is in fact the lowest end of this range of estimates that has been mostly widely used in policy analyses in the U.S. involving gasoline taxes or other incentives to reduce demand. Even the high end of this range, however, is probably too low; as we have said elsewhere in this

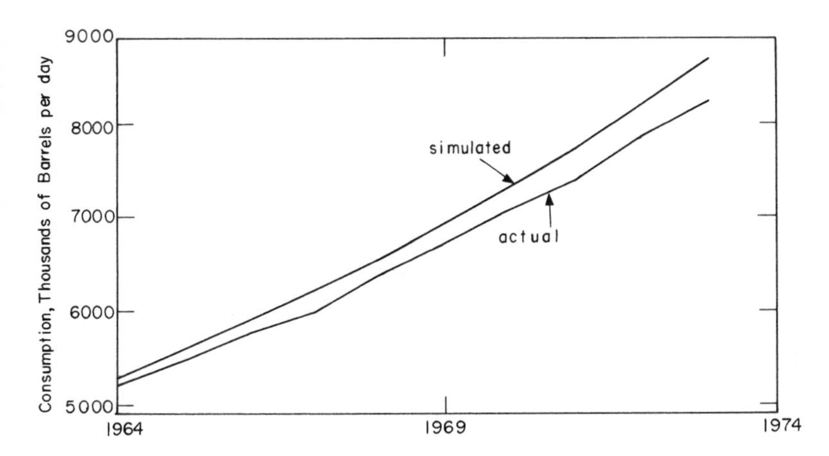


**Figure 6.4**
Historical simulation, nine countries

book, the use of data for a single country is more likely to elicit short- or intermediate-run elasticities. This should be particularly true for gasoline demand since, as we have seen from our model, a long period of time must elapse (fifteen to twenty years) for the full adjustment to take place to long-run equilibrium. Indeed, Adams, Graham, and Griffin (1), working with international data (for twenty OECD countries in the single year 1969) obtained a price elasticity estimate of −0.92, which is much closer to ours.[6] It would therefore appear that this elasticity is in the vicinity of at least −1, so that higher gasoline prices could be a very effective means of reducing consumption, although a number of years would have to pass for the effects of price increases to take place.

One means of evaluating the model as a whole is to simulate the model over some historical time period (we use the period 1964 to 1973) and compare predicted gasoline consumption with its actual values. The results of such a simulation are shown in figure 6.4 for total gasoline consumption in nine countries (Canada and Switzerland are omitted because data for some of the exogenous variables is not available for the years prior to 1968), and in figure 6.5 for

6. As for the long-run income elasticity of gasoline demand, our results are within the range of other studies. Houthakker et al. (93) obtained an estimate of 0.98, Ramsey, Rasche, and Allen (144) obtained 1.15, and Adams, Graham, and Griffin (1) 0.54.

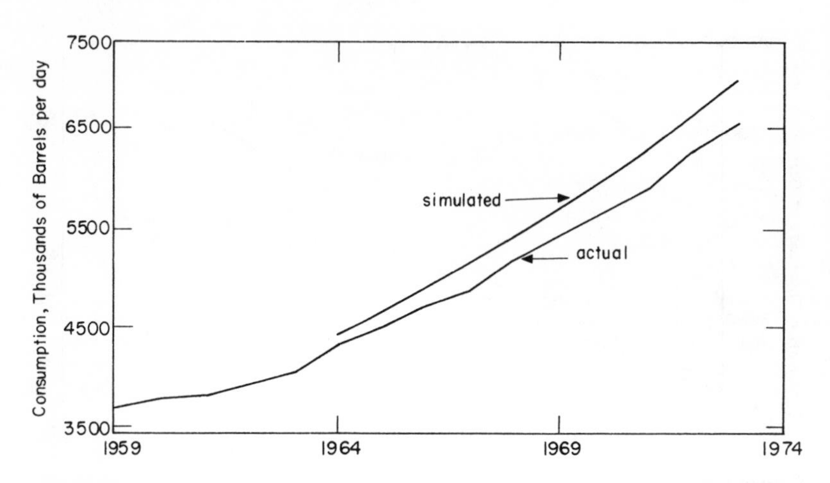


**Figure 6.5**
Historical simulation, the U.S.

gasoline consumption in the U.S. Observe that the model overpredicts gasoline consumption (largely because of an overprediction of traffic volume per car), but the prediction error never exceeds 7 percent.

Our objective throughout this book has been to study the structure of energy demand, and not to forecast future energy use. It is, however, interesting to ask whether the large long-run price elasticities that we have found for gasoline imply that gasoline consumption will stop growing or will even fall in the future given that prices have already increased significantly. We therefore show in table 6.5 the projections of gasoline consumption corresponding to the base case simulation described earlier. Recall that in the simulation the real price of gasoline in each country increases from its actual 1976 value at a rate of 4 percent per year. Observe from table 6.5 that despite actual past and projected future increases in the price of gasoline, consumption is still projected to grow over the next decade or so. For the total of the eleven countries, consumption grows by 2.5 percent per year between 1980 and 1985, and about 2 percent per year between 1985 and 1990. This increase occurs because of GDP growth and because the model predicts that the stock of cars will rise (that is, new registrations will exceed retirements) even with a fixed GDP. The rate of increase is, however, smaller than it has been in the past, and smaller than rates of increase forecasted by such groups as the OECD (125).

## 6.2 The Demands for Other Fuels

We turn now to the estimation of demand elasticities for other fuels used in the transportation sector. We use simple log-linear equations with a Koyck lag adjustment (that is, with a lagged dependent variable) to explain the dynamic adjustment of demand to changes in price or income. The equation is of the form

$$\log q_{ijt} = \alpha_{ij} + \beta_i \log y_{jt} + \gamma_i \log P_{ijt} + \lambda_i \log q_{ij,t-1} + \epsilon_{ijt}, \qquad (6.7)$$

where $q_{ijt}$ is per capita consumption on fuel $i$ in country $j$ at time $t$, $y_{jt}$ is per capita GDP in country $j$, and $p_{ijt}$ is the real retail price of fuel $i$. The long-run income and price elasticities of demand implied by this equation are $\beta_i/(1 - \lambda_i)$ and $\gamma_i/(1 - \lambda_i)$, respectively.

Equation (6.7) is estimated for each of four fuels: aviation gasoline (used to power propeller aircraft), jet fuel, diesel fuel, and motor gasoline. (The last fuel is included to provide a comparison of elasticities generated from equation (6.7) with those generated by the model presented in the last section.) For each fuel the equation is estimated for each of two sets of countries: a group of twelve European countries (Austria, Belgium, Denmark, France, Ireland, Italy, the Netherlands, Norway, Sweden, Switzerland, the U.K., and West Germany), and the U.S. and Canada. The data cover the period 1955 to 1973. In all cases the equation is estimated using ordinary least squares.

Estimation results are shown for the European countries in table 6.6, and for the U.S. and Canada in table 6.7 (test statistics are in parentheses). We turn first to the results in table 6.6 for the European countries. Observe first that equation 5 for motor gasoline gives long-run price and income elasticities of −1.61 and 0.66 respectively, which are consistent with the elasticities generated by the multi-equation model of the last section (−1.31 and 0.84, although with a median lag about twice as long as in the log-linear equation). If the cross-sectional variation in the prices and quantities of the other fuels is as great as it is for gasoline, we might expect to obtain reasonable estimates of long-run elasticities.

Note that the estimated long-run price elasticity of diesel fuel demand is much smaller than for gasoline (though with a value of −0.62 still larger than the consensus estimates often used for policy analysis). We would expect this since for the commercial vehicles that burn diesel fuel, fuel costs are a much smaller proportion of total cost (capital plus operating) than is the case for private cars, first because of the lower cost of diesel fuel combined with the higher cost of diesel engines, and second because of scale effects.

Long-run price elasticities for jet fuel and aviation gasoline are still lower (−0.3 to −0.4), while the long-run GDP elasticities are all above 2. We would expect the price elasticities to be low since the demand for these fuels depends basically on the demand for air transportation—and with fuel costs only a small part of the total cost of air transportation. This latter demand should not change much when fuel prices increase. The demand for air transportation should, however, be highly dependent on income or GDP, so that it is reasonable to observe a GDP elasticity of fuel demand greater than 2. (The value of 6.26 for aviation gasoline, however, is unrealistic and is not statistically significant at the 5 percent level.) Note that the equation for aviation gasoline contains a time trend variable which appears with a negative and highly significant coefficient. This accounts for the fact that over the years there has been a shift from propeller planes to jets that has been unrelated to relative fuel prices. Note also that the jet fuel equation has been estimated both with and without the lagged dependent variable because the inclusion of that variable yields a price coefficient that is not significant at the 5 percent level. The price elasticities for the two equations, however, are about the same in the long run.

We turn next to table 6.7 for the U.S. and Canada. Here we find much larger price elasticities for aviation gasoline, jet fuel, and diesel fuel, but smaller ones for motor gasoline. The extremely large price and GDP elasticities for aviation gasoline cannot be explained, and in any case are based on estimated coefficients that are not statistically significant. The GDP elasticities for jet fuel appear reasonable, but the price elasticities seem much too large. In equation 2 the estimate is based on a statistically insignificant coefficient, and in equation 3 we can only assume that the estimate is an artifact resulting from a spurious correlation.

Equations for diesel fuel are estimated both with and without the lagged dependent variable. The resulting price and GDP elasticities for the two equations are almost identical, and in both cases are based on highly significant coefficients. The GDP elasticity of about 1 is reasonable, while the price elasticity around −1 seems somewhat large, although plausible. The two equations for motor gasoline, however, both give very low price elasticities, probably because the small amount of cross-sectional variation resulting from pooling only two countries is insufficient to elicit the true long-run elasticity. This in turn makes the results for diesel fuel even harder to explain.

The most meaningful results would appear to be those in table 6.6 for the European countries. These results confirm the large long-run price elasticity for motor gasoline found in the last section, show diesel fuel to be less price elastic than motor gasoline, and jet fuel and aviation gasoline to be still less price elastic but highly income elastic. The elasticities reported in the table, however, should be viewed as general indicators rather than exact estimates, since they are based on a crude and extremely simple model of demand.

**Table 6.1**
Gasoline demand model, intercept coefficients for estimated equations

| Equation | Belgium | Canada | France | Italy | Netherlands | Norway |
|---|---|---|---|---|---|---|
| (6.1) | 25,147 | 23,002 | 25,031 | 25,727 | 24,204 | 21,093 |
| | (4.10) | (3.90) | (4.00) | (4.22) | (4.01) | (3.70) |
| (6.2) | 0.108 | 0.098 | 0.089 | 0.061 | 0.078 | 0.070 |
| | (7.04) | (6.50) | (5.95) | (4.25) | (4.83) | (5.00) |
| (6.3) | −0.899 | −0.817 | −0.872 | −0.824 | −0.851 | −0.875 |
| | (−2.48) | (−2.34) | (−2.38) | (−2.30) | (−2.44) | (−2.42) |
| (6.4) | 1.12 | 1.06 | 1.12 | 1.07 | 1.07 | 1.12 |
| | (2.32) | (2.36) | (2.28) | (2.14) | (2.20) | (2.31) |
| (6.4a) | 2.13 | 1.98 | 2.12 | 2.02 | 2.03 | 2.12 |
| | (2.08) | (2.10) | (2.06) | (1.93) | (1.98) | (2.08) |

| Sweden | Switzerland | U.K. | U.S. | West Germany | $R^2$ | $F$ |
|---|---|---|---|---|---|---|
| 22,753 | 21,690 | 22,208 | 25,977 | 23,693 | 0.936 | 130 |
| (3.83) | (4.01) | (3.90) | (4.11) | (4.03) | | |
| 0.092 | 0.099 | 0.088 | 0.104 | 0.096 | 0.519 | 12.7 |
| (6.69) | (6.86) | (6.25) | (7.22) | (6.65) | | |
| −0.869 | −0.876 | −0.857 | −0.870 | −0.876 | 0.980 | 377 |
| (−2.41) | (−2.44) | (−2.41) | (−2.42) | (−2.51) | | |
| 1.11 | 1.10 | 1.10 | 1.08 | 1.10 | 0.988 | 599 |
| (2.31) | (2.37) | (2.27) | (2.41) | (2.30) | | |
| 2.11 | 2.08 | 2.09 | 2.01 | 2.11 | 0.983 | 415 |
| (2.08) | (2.12) | (2.05) | (2.13) | (2.10) | | |

**Table 6.2**
Elasticities of endogenous variables[a]

| | with respect to: | | | | | |
|---|---|---|---|---|---|---|
| | *PC* | | *PG* | | *GDP/POP* | |
| Elasticity of: | Short run | Long run | Short run | Long run | Short run | Long run |
| (6.1) New registrations | −0.32 | −0.78 | −0.26 | −0.64 | 0.12 | 0.30 |
| (6.2) Depreciation rate | −0.71 | −0.71 | — | — | — | — |
| (6.3) Traffic volume per car | — | — | — | — | 0.06 | 0.66 |
| (6.4) Efficiency | — | — | 0.11 | 1.43 | — | — |
| (6.4a) Efficiency | — | — | 0.12 | 2.63 | — | — |

[a] Evaluated at point of means.

**Table 6.3**
Price elasticities of gasoline demand[a]

| Country | Year 1 | 2 | 3 | 4 | 5 | 10 | 15 | 20 | 25 |
|---|---|---|---|---|---|---|---|---|---|
| Belgium | 0.124 | 0.248 | 0.366 | 0.478 | 0.581 | 0.981 | 1.24 | 1.40 | 1.51 |
| Canada | 0.110 | 0.214 | 0.310 | 0.399 | 0.481 | 0.800 | 1.01 | 1.14 | 1.23 |
| France | 0.126 | 0.252 | 0.374 | 0.489 | 0.596 | 1.02 | 1.30 | 1.48 | 1.60 |
| Italy | 0.051 | 0.117 | 0.188 | 0.259 | 0.328 | 0.621 | 0.838 | 1.00 | 1.13 |
| Netherlands | 0.121 | 0.242 | 0.357 | 0.465 | 0.565 | 0.958 | 1.22 | 1.39 | 1.51 |
| Norway | 0.137 | 0.280 | 0.421 | 0.556 | 0.683 | 1.19 | 1.54 | 1.77 | 1.94 |
| Sweden | 0.119 | 0.237 | 0.349 | 0.454 | 0.551 | 0.935 | 1.19 | 1.35 | 1.47 |
| Switzerland | 0.120 | 0.240 | 0.354 | 0.460 | 0.560 | 0.948 | 1.20 | 1.37 | 1.48 |
| U.K. | 0.131 | 0.266 | 0.399 | 0.524 | 0.642 | 1.11 | 1.42 | 1.63 | 1.77 |
| U.S. | 0.111 | 0.217 | 0.315 | 0.406 | 0.490 | 0.818 | 1.03 | 1.17 | 1.26 |
| West Germany | 0.117 | 0.230 | 0.338 | 0.438 | 0.530 | 0.891 | 1.13 | 1.28 | 1.38 |
| Total: | 0.111 | 0.219 | 0.320 | 0.414 | 0.501 | 0.842 | 1.06 | 1.21 | 1.31 |

[a] All of these elasticities are negative, but the minus signs have been left off.

**Table 6.4**
Income Elasticities of gasoline demand

| Country | Year 1 | 2 | 3 | 4 | 5 | 10 | 15 | 20 | 25 |
|---|---|---|---|---|---|---|---|---|---|
| Belgium | 0.066 | 0.130 | 0.192 | 0.251 | 0.305 | 0.516 | 0.651 | 0.736 | 0.790 |
| Canada | 0.067 | 0.133 | 0.197 | 0.257 | 0.313 | 0.534 | 0.675 | 0.765 | 0.823 |
| France | 0.067 | 0.136 | 0.201 | 0.264 | 0.322 | 0.554 | 0.706 | 0.806 | 0.871 |
| Italy | 0.065 | 0.129 | 0.190 | 0.248 | 0.303 | 0.521 | 0.668 | 0.768 | 0.837 |
| Netherlands | 0.066 | 0.132 | 0.196 | 0.256 | 0.312 | 0.533 | 0.678 | 0.774 | 0.837 |
| Norway | 0.070 | 0.143 | 0.214 | 0.282 | 0.346 | 0.607 | 0.784 | 0.905 | 0.988 |
| Sweden | 0.069 | 0.139 | 0.207 | 0.271 | 0.332 | 0.575 | 0.735 | 0.840 | 0.909 |
| Switzerland | 0.069 | 0.138 | 0.205 | 0.269 | 0.329 | 0.567 | 0.722 | 0.822 | 0.888 |
| U.K. | 0.068 | 0.137 | 0.204 | 0.267 | 0.327 | 0.566 | 0.724 | 0.829 | 0.898 |
| U.S. | 0.067 | 0.133 | 0.198 | 0.259 | 0.316 | 0.540 | 0.685 | 0.778 | 0.837 |
| West Germany | 0.067 | 0.134 | 0.198 | 0.259 | 0.316 | 0.539 | 0.684 | 0.777 | 0.838 |
| Total: | 0.067 | 0.134 | 0.198 | 0.259 | 0.316 | 0.541 | 0.686 | 0.780 | 0.841 |

**Table 6.5**
Projected gasoline consumption (thousands of barrels per day)

| | 1980 | 1985 | 1990 |
|---|---|---|---|
| Belgium | 76 | 82 | 86 |
| Canada | 944 | 1,189 | 1,397 |
| France | 572 | 704 | 809 |
| Italy | 359 | 434 | 510 |
| Netherlands | 123 | 141 | 155 |
| Norway | 29 | 31 | 32 |
| Sweden | 111 | 123 | 132 |
| Switzerland | 77 | 82 | 86 |
| U.K. | 462 | 497 | 521 |
| U.S. | 8,050 | 8,868 | 9,570 |
| West Germany | 744 | 904 | 1,037 |
| Total: | 11,548 | 13,056 | 14,355 |

**Table 6.6**
Demand estimates for Europe[a]

| Dependent variable | Price | GDP | Time | Lagged dependent |
|---|---|---|---|---|
| 1. Aviation gasoline | −0.413 | 6.26 | −0.099 | — |
| | (−1.01) | (1.34) | (−6.33) | |
| 2. Jet fuel | −0.094 | 0.586 | — | 0.779 |
| | (−1.64) | (3.02) | | (21.80) |
| 3. Jet fuel | −3.25 | 3.80 | — | — |
| | (−3.14) | (16.50) | | |
| 4. Diesel fuel | −0.075 | 0.096 | — | 0.878 |
| | (−1.93) | (1.85) | | (46.90) |
| 5. Motor gasoline | −0.214 | 0.088 | — | 0.867 |
| | (−5.91) | (2.46) | | (48.59) |

[a] Country intercept parameters not shown.
[b] Time required to reach one-half of total adjustment in demand = $\log(0.5)/\log \lambda$.

| Long-run elasticities | | Median | | |
|---|---|---|---|---|
| Price | GDP | Lag[b] | $R^2$ | $F$ |
| −0.41 | 6.26 | — | 0.820 | 68 |
| −0.43 | 2.65 | 2.78 | 0.941 | 236 |
| −0.33 | 3.80 | — | 0.806 | 65 |
| −0.62 | 0.79 | 5.33 | 0.988 | 1267 |
| −1.61 | 0.66 | 4.86 | 0.995 | 3116 |

**Table 6.7**
Demand estimates for U.S. and Canada

| Dependent variable | Price | GDP | Time | Lagged dependent |
|---|---|---|---|---|
| 1. Aviation gasoline | −0.594 | 1.23 | −0.069 | 0.801 |
| | (−1.43) | (1.34) | (−1.68) | (7.77) |
| 2. Jet fuel | −0.571 | 0.831 | — | 0.603 |
| | (−1.46) | (2.92) | | (6.89) |
| 3. Jet fuel | −1.82 | 2.25 | — | — |
| | (−3.38) | (7.38) | | |
| 4. Diesel fuel | −0.440 | 0.373 | — | 0.597 |
| | (−3.11) | (3.63) | | (9.39) |
| 5. Diesel fuel | −1.06 | 1.02 | — | — |
| | (−4.52) | (7.10) | | |
| 6. Motor gasoline | −0.032 | 0.355 | — | 0.686 |
| | (−0.45) | (5.69) | | (9.61) |
| 7. Motor gasoline | −0.381 | 0.860 | — | — |
| | (−3.24) | (13.3) | | |

| Constants | | Long-run elasticities | | Median | | |
|---|---|---|---|---|---|---|
| U.S. | Canada | Price | GDP | Lag | $R^2$ | $F$ |
| 5.77 (0.84) | 2.95 (0.44) | −2.98 | 6.18 | 3.12 | 0.957 | 144 |
| 0.994 (0.22) | 0.981 (0.21) | −1.44 | 2.09 | 1.37 | 0.982 | 429 |
| 4.40 (0.63) | 4.45 (0.63) | −1.82 | 2.25 | — | 0.954 | 230 |
| 4.06 (2.21) | 3.95 (2.22) | −1.09 | 0.93 | 1.34 | 0.986 | 564 |
| 8.98 (2.71) | 8.70 (2.70) | −1.06 | 1.02 | — | 0.947 | 202 |
| 0.283 (0.28) | 0.277 (0.27) | −0.10 | 1.13 | 1.84 | 0.996 | 2324 |
| 5.75 (3.57) | 5.67 (3.53) | −0.38 | 0.86 | — | 0.987 | 832 |

# 7 Energy Demand in the Developing Countries

The less developed countries (LDC) account for a small but significant portion of non-Communist world energy demand. In 1977 primary energy consumption in Greece, Portugal, Spain, Turkey, and all of the non-OECD non-Communist countries amounted to about 10.9 million Tcals (43.3 Quads), while consumption in the remaining OECD countries was about 34.7 million Tcals (138.4 Quads).[1] Thus this first group of countries, which we will classify as LDC, accounted for about 24 percent of primary energy consumption in 1977.[2]

Although LDC energy consumption is not large, particularly when viewed on a per capita basis, the argument has been raised that energy use in these countries is likely to rise disproportionately as their economies grow, so that they will represent a major source of increased demands for energy in the future, and will account for a growing share of world energy consumption. In fact, this share has not grown very much in the past; in 1967, for example, these four developing countries consumed 6.3 million Tcals (25.3 Quads) of primary energy but had a total share of 22 percent, only 2 percentage points below the 1977 value. On the other hand, there is some evidence that in the developing countries price elasticities of energy demand tend to be smaller, but income elasticities tend to be larger than in the developed countries. As long as economic growth rates do not fall, this could mean that over the next decade, with energy prices much higher than in the past, the percentage share of LDC energy demand will grow. Unfortunately, economic growth rates in these countries may fall somewhat; energy price increases that have occurred already may have a significant depressive effect on the

1. This includes only commercial energy, that is, oil, natural gas, coal, water power, and nuclear, and does not include such noncommercial energy sources as firewood and animal waste. Source: *BP Statistical Review of the World Oil Industry,* 1977.
2. Obviously this classification is somewhat arbitrary. Greece, for example, has a much higher level of per capita GNP than, say, India. Our objective here, however, is not to analyze the structures of these economies in any detail, so for purposes of rough comparison we loosely define Greece, Portugal, Spain, Turkey, and the non-OECD countries as "developing" and the remaining OECD countries as "developed."

economies of the developing countries, which could reduce their share of energy use.[3]

Clearly a better understanding of the structure of energy demand in the developing countries is needed if we are to be able to assess the future role of these countries in world energy markets, or to determine the probable effects of higher energy prices on their economies. Unfortunately, it is much more difficult to study the characteristics of energy demand in these countries on an empirical basis. The first and most serious problem is that for most of the countries there is little good data available, particularly for the retail or wholesale prices of various fuels. Thus, even where fuel quantity data is available (which it is at least on an aggregated level for a number of developing countries), it is impossible to estimate econometric models specified to explain the role of prices. In addition, for some of the poorer countries a large fraction of energy use consists of noncommercial fuels such as wood and animal waste for which meaningful market prices do not even exist.

A second problem is that demand models based on the notion of industrial and residential consumers facing market prices and making cost-minimizing or utility-maximizing decisions may simply not apply for many LDC's. In the industrial sector, for example, both government and private enterprises may, because of operating constraints or management objectives other than profit maximization, be less likely to base energy consumption decisions on cost minimization. In the residential sector, consumption decisions may be much more a function of supply availability than price; commercial fuels, and in particular electricity, are simply not available in some regions.

A third problem is that for many LDC's the structure of the economy may be changing in a way that makes the structure of energy demand itself change rapidly. Examples of this include rapid urban migration, industrial modernization, and the sudden introduction of commercial energy supplies into particular regions of a

3. We discuss the impact of higher energy prices on aggregate economic output in the next chapter, but largely in the context of the developed countries. Here we just note that for the LDC's the impact might be much greater, particularly if more expensive imported energy squeezed the already limited ability of these countries to import capital goods.

country. These changes could bias parameter estimation for models based on a static structure of energy demand.

Because of these problems we have attempted to obtain only some very limited empirical estimates of energy demand elasticities, and these only for some of the more advanced developing countries for which the necessary data is available. We will, however, offer some qualitative arguments for why (and how) energy demand elasticities in the developing countries should differ from those in the developed countries, and we will use our empirical estimates to test these arguments. In addition, we will briefly discuss the role of the developing countries in world energy use—and the importance of energy to the economic growth of these countries.

### 7.1 Energy Use in the Developing vs. the Developed Countries

Predictions that energy demand will grow more rapidly in the developing than in the developed countries over the next two decades are often based on the argument that income or GNP elasticities of energy demand are higher in the developing countries. If we examine the ratio of per capita energy consumption to per capita income as a crude proxy for the income elasticity of energy demand, we indeed observe much larger numbers for the developing countries. Also, to the extent that estimated income elasticities of energy demand are available for a cross section of both developed and developing countries, they show that, as per capita income rises, the income elasticity of energy demand falls. (This is also true within the group of developed countries alone.)

The argument that income elasticities are higher would apply largely to the residential and transportation sectors, which, for the developing countries, represent a greater fraction of total demand. As incomes rise, additional expenditures are not allocated proportionately to larger homes or to more heat or light in existing homes. Thus expenditures on energy, like those on food and shelter, represent a large fraction of the consumer budget when incomes are low but a smaller fraction as incomes rise.

There are other reasons that we might expect energy demand to grow more rapidly in the developing countries. One of these is that a large proportion of energy consumption is from noncommercial sources, of which the major ones are firewood, dung, and agricultural waste. This is particularly true in the poorer countries; in India, for example, nearly half of the Btu's consumed in recent years came from such sources. The share of noncommercial energy, however, is falling. In India, noncommercial energy accounted for two-thirds of the energy consumed in 1955, but, as Parikh (126) shows, this fraction dropped steadily to one-half by 1968, and reached 40 percent by 1977. This means that for any percentage growth rate of total energy use, there will be a greater percentage growth rate of commercial energy demand (coal, oil, and electricity).

The fuel mix in the developing countries is also changing, with the proportions of oil and electricity growing, largely at the expense of noncommercial energy sources and to some extent at the expense of coal. The share of electricity in total energy use is growing particularly rapidly; in India it accounted for only 4 percent of final energy consumption in 1955, but rose to 13 percent by 1970. Given the large Btu requirement of primary energy to generate a Btu-equivalent of electricity, if the percentage share of electricity continues to grow, it will mean an even faster growth rate of primary fuel requirements.[4]

Urban migration may also contribute to a more rapid growth rate of household energy demand in the developing countries. In rural areas, where a greater proportion of energy use is noncommercial, household energy demand is likely to be very price inelastic, but also relatively income inelastic. Parikh (126) observes that in rural areas of India the use of firewood and other noncommercial fuels does not depend much on total household consumption levels. In urban areas, however, household demand is likely to be much more income elastic since a larger share of any increase in consumption expenditures will be allocated to housing and the associated use of energy. This means that urban migration could increase total energy demand at any particular level of income.

The factors discussed above would contribute to a higher income elasticity of energy demand in the developing countries. It is also likely that in the residential sector, and also to some extent in the

4. In India, the share of oil grew from 15 percent in 1955 to 25 percent in 1970. The share of coal fell slightly, from 15 percent in 1955 to 14 percent in 1970. The noncommercial energy share fell from 66 percent in 1955 to 48 percent in 1970.

transportation sector, price elasticities of energy demand will be lower. the reason is simply that at low levels of income most energy is consumed as a necessity, while as incomes grow, the additional use of energy becomes more discretionary, allowing for greater substitution away from energy if prices rise.

It is important to recognize that low-price elasticities and high-income elasticities in the residential sector do not necessarily mean that the LDC's will consume much more energy as their per capita incomes increase. It is quite possible that if energy prices continue to rise rapidly, energy-importing LDC's will deal with growing trade deficits by imposing stiff taxes and and/or quotas on fuels, so that consumers in those countries may face very large price increases. In addition, future energy use in many countries may be much more determined by availabilities of supplies than by the structure of demand—which to some extent has already been the case.

In the industrial sector the ratio of energy use to GDP, and the GDP elasticity of energy demand, have been historically lower in the developing countries (as has been the share of energy consumed in the industrial sector). This is largely due to the fact that much of the growth in output has occurred in agriculture and light industry where energy requirements have been small, and where labor—a relatively cheap factor—can easily be substituted for energy. However, if economic development means that the industrial structure of the LDC's will become more like the industrial structure of the developed countries, then in fact the long-run GDP elasticity of energy demand may be considerably higher than we have thought to be the case.

The LDC's have always lagged the developed countries in the production of those basic materials that require large amounts of energy (for example, steel, basic metals, paper, pulp, and fertilizer).[5] Most LDC's import most of the energy-intensive materials

5. This is discussed in some detail by Strout (152, 153). He points out that the production of energy-intensive materials accounts for one-quarter of all energy demand in the U.S., and over one-half of all energy use in such countries as Japan, Austria, Norway, Finland, Sweden, Belgium, and Canada. This is also supported by the recent analysis in Darmstadter, Dunkerley, and Alterman (48). The U.S. is actually a net importer (the world's largest) of energy-intensive materials, despite its large production of these materials.

they use, and this reduces their consumption of energy as a fraction of GDP. To the extent that economic growth means an increase in the production of energy-intensive materials (as a substitute for higher-priced imports), industrial demand for energy will grow more rapidly than in the developed countries. So far, however, there is no indication that this will be the case. Even as real energy prices fell during the 1960s, most of the developing countries made little progress in reducing the gap between their consumption and production of energy-intensive materials. Higher energy prices may make this gap widen even more, with the industrial structure of the developing countries evolving towards reduced production and consumption of energy-intensive materials.

We could therefore expect the GDP elasticity of industrial energy demand in most of the developing countries to be below the value of 0.80 to 0.85 that we have estimated for some of the industrialized countries. Also, there is likely to be greater substitutability between energy and capital (and labor) for many of the developing countries, again because of the smaller share of energy-intensive materials in production. This would mean that the price elasticity of the industrial demand for energy should be larger for the LDC's, as labor and capital can be substituted for energy to a greater extent.

We now turn to some empirical estimates of demand elasticities for various petroleum products in several developing countries. Because of data limitations, the models that we estimate are extremely simple and are applied to only a few countries, so that the results are of limited value for predictive purposes. On the other hand, they provide at least some means of testing whether demand elasticities in the developing and developed countries differ in the ways that we expect.

## 7.2 Some Empirical Estimates of Demand Elasticities

We present here estimated equations for the demands for several petroleum products. Because of the data limitations, these equations are estimated for only five countries, which are pooled in two groups. The first group consists of Greece, Spain, and Turkey and the second of Brazil and Mexico. (This division is based on a greater homogeneity of demand structure within each group, as well as differences in the sources and aggregation of the data.)

Equations for Greece, Spain, and Turkey are estimated for each of six petroleum products: light fuel oil, which is used almost exclusively in the residential sector, heavy (residual) fuel oil, which is consumed in the industrial sector, motor gasoline, diesel fuel, kerosene, much of which is consumed in the residential sector, and "other petroleum products," which consist largely of chemical feedstocks and fuel oils used for electricity generation. The quantity data for these countries were obtained from the OECD's *Energy Statistics.* For Brazil and Mexico, however, it was necessary to use the more aggregated data from the U.N.'s *World Energy Supplies.* For these two countries equations are estimated for four fuel categories: fuel oils (light and heavy), gasolines (motor and aviation), kerosene, and liquified petroleum gas (LPG), some of which is consumed by all sectors but most by the residential sector.

We estimate simple log-linear models with a Koyck lag to explain the dynamic adjustment of demand to changes in income or price. The equation is of the form.

$$\log q_{ijt} = \alpha_{ij} + \beta_i \log y_{jt} + \gamma_i \log p_{ijt} + \lambda_i \log q_{ij,t-1} + \epsilon_{ijt}, \qquad (7.1)$$

where $q_{ijt}$ is per capita consumption of fuel $i$ in country $j$ at time $t$, $y_{jt}$ is per capita gross domestic product (GDP) in country $j$, and $p_{ijt}$ is the price of fuel $i$. Note that the long-run income and price elasticities of demand implied by this equation are $\beta_i/(1 - \lambda_i)$ and $\gamma_i/(1 - \lambda_i)$, respectively.

In some cases additional variables are included in equation (7.1). These can include temperature (measured as the average temperature over the five winter months November to March in the major cities of each country), the prices of other fuels (electricity, natural gas, and coal) or an aggregate price of energy (obtained from the residential or industrial translog aggregator equations estimated in chapters 4 and 5 applied to the fuel prices in these countries). All of the prices are retail, and for each country are measured in real local currency units per Tcal of energy.

Our data for Greece, Turkey, and Spain cover the twenty years 1955 through 1974. Data for Brazil and Mexico span the periods 1954 to 1974 and 1960 to 1974 respectively. In each case, equation (7.1) is estimated using ordinary least squares.

Estimation results for Greece, Spain, and Turkey are shown in table 7.1, and for Brazil and Mexico in table 7.2[6] (test statistics are in parentheses). Let us turn first to the results for Greece, Spain, and Turkey. Observe that these results are quite consistent with the arguments raised in the last section. The demand for light fuel oil, which is used in the residential sector, appears to be price inelastic but highly income elastic as expected. (The price term is positive and insignificant in equation 1, and was dropped from equation 2.) The same is true for motor gasoline and diesel fuel. Price elasticities for gasolines and light fuel oil range from 0 to −0.4, and income elasticities range from about 1.5 to about 3.2. Heavy fuel oil, on the other hand, is highly price elastic but less income elastic. This is again consistent with our expectation that energy demand in the industrial sector should be more price elastic in the developing countries because of a greater ability to substitute low-priced labor.

The demand for kerosene is price inelastic (the price term is positive and insignificant in equation 8 and is dropped in equation 9), and has an income elasticity that is negative. This is consistent with the fact that as a fuel for residential use, kerosene is an inferior good, so that its use for home heating and cooking decreases as incomes rise, and other fuels can be substituted in its place. (Over the decade 1963 to 1973 the consumption of kerosene as a share of total petroleum products dropped from 3.4 to 0.4 percent in Greece, from 5.0 to 0.7 percent in Spain, and from 18.1 to 4.9 percent in Turkey.)

Finally, an equation is estimated for other petroleum products. This includes chemical feedstocks and fuel oils used for the generation of electricity. The price of electricity is included as an additional variable in this equation (as the price of electricity increases, its use should decrease, thereby decreasing the demand for this petroleum product category), but its coefficient is statistically insignificant. Note that the own-price elasticity is relatively small (there is less room for substitution), and the income elasticity is large.

We turn now to the estimates for Brazil and Mexico. Unfortunately, the available data aggregate light and heavy fuel oils to-

6. These results are described in more detail in Heide (82).

gether, so that equation 1 in table 7.2 applies to demands in both the residential and industrial sectors. The low estimated price elasticity in that equation is consistent with our expectations about the residential sector, but not the industrial sector. Gasolines, on the other hand, are used only in the transportation sector, and the estimates in equation 2 are consistent with our expectations; demand is relatively price inelastic, but income elastic. (The estimated price elasticity of gasoline demand is −0.55 as compared to the estimate of about −1.3 obtained for the developed countries in the last chapter, while the income elasticity is 1.22 as compared to 0.8 for the developed countries.)

Kerosene, as explained earlier, is used largely in the residential sector. Equation 3 shows a price elasticity that is very small, and an income elasticity insignificantly different from zero, which is consistent with a shift to alternative fuels as incomes rise. Finally, the demand for liquified petroleum gases (LPG), which is used largely in the residential sector, shows a moderate price elasticity, but an income elasticity that is large, as expected.

It would be useful to compare these elasticity estimates with those for some of the industrialized countries. Because the residential and industrial energy demand elasticities estimated in chapters 4 and 5 are based on models very different in structure from those estimated here, it might be misleading to make the comparison with the elasticity estimates obtained in those earlier chapters. Instead, we have estimated equation (7.1) for light and heavy fuel oil consumption in two groups of countries: the U.S. and Canada, and twelve European countries.[7]

Estimates of equation (7.1) for the industrialized countries are shown in table 7.3, and can be compared with the corresponding estimates in table 7.1. Observe that price elasticities of demand for light fuel oil are around −0.8 to −1.2, while the income elasticities are between 0.6 and 1.5. Again, we see that the residential demand for oil is much less price elastic and much more income elastic in the developing countries. Price elasticities for heavy fuel oil range

7. The European countries include Austria, Belgium, Denmark, France, Ireland, Italy, the Netherlands, Norway, Sweden, Switzerland, the U.K., and West Germany. All of these countries are pooled together in estimating the equation.

from −0.6 to −1.0, while income elasticities range from about 0 to 1.6. Thus, although we can draw no conclusion about income elasticities, the price elasticity of industrial fuel oil demand is clearly higher in the developing countries. Again this supports the qualitative arguments made in the last section.

## 7.3 The Future of the Developing Countries as Energy Consumers

One of the difficulties in predicting the future energy consumption of the LDC's and the role that they will have in world energy markets is that their energy use is highly dependent on their economic growth rates. We have seen that income elasticities of energy demand tend to be high for the developing countries, particularly in the residential and transportation sectors, while price elasticities are low, so that rapid economic growth in these countries could contribute to large increases in energy demand. The problem is that there is considerable uncertainty over just what LDC growth rates will be during the next decade or two. In addition, it is not clear to what extent LDC governments will be forced to take measures to limit energy consumption, even if growth rates are high.

There is little doubt that the sharp increase in energy prices that occurred in 1973 to 1974 has had, and will continue to have, an adverse impact on the economic growth prospects of the developing countries. Aside from the oil exporting countries and a small number of additional countries that have the capacity to develop new domestic sources of energy, most of the developing countries will continue to remain dependent on expensive imported oil for some time, and will reduce this dependence only slowly.[8]

As Tims (159) points out, for many LDC's it is unlikely that the higher cost of energy, and imported oil in particular, can be offset either by a major reduction in the use of energy, or by substitution from imported oil to other lower-priced fuels. Substitution possibilities are limited in the short and medium term, particularly in the

8. The OPEC countries are of course not typical of most LDC's in terms of future potential energy use. Energy demand in the OPEC countries is expected to grow very rapidly. In terms of oil alone, the International Energy Agency expected demand to double between 1974 and 1985. See Hamilton (80).

electric power sector where long lead times are needed for new investments. And as we can infer from the low price elasticities in the residential and transportation sectors, the ability to reduce energy use without seriously affecting standards of living is also limited. Some countries may develop investment programs for the future production of domestic energy, but such investments will be highly capital-intensive and will require substantial imports of machinery and equipment, which will mean a reduction in other investment programs. For those countries that do not undertake investment programs in domestic production, the worsening terms of trade will mean a reduced ability to import capital goods in general. In either case nonenergy capital accumulation will proceed more slowly, and growth rates will fall.

The magnitude of this problem becomes clear if we note that the increase in oil prices raised the aggregate current account deficit of the non-OPEC LDC's from $10 billion in 1973 to $35 billion in 1975 (it later fell to more manageable levels—$25 billion in 1976 and about $23 billion in 1977), with an increase in total obligations from about $70 billion in 1973 to about $140 billion in 1976 and $160 billion in 1977.[9] (These are current dollar figures; in real terms total obligations increased by some 60 percent over the period.) The impact on GNP growth rates has so far been limited—a 4 percent drop in the aggregate GNP of these countries occurred in the worst year, 1975. But this limited impact on GNP occurred at the cost of a major increase in external indebtedness, and this has created a debt burden which, particularly for the poorest countries, may severely limit their ability to borrow in the coming years, thus restraining GNP growth in the future. The point here is that the full impact on the LDC's of energy price increases that have occurred in the past has probably not yet taken place.

It is much more difficult to determine quantitatively the impact of an increase in energy prices on the economic output of the developing countries than it is for the developed countries. As we will see in the next chapter, for the developed countries the impact can at least be partially assessed to the extent that one has an accurate quantitative model of the structure of industrial production, and in particular of the substitutability of energy with other

9. *Petroeconomic File,* April 1978, no. 15.

factors. For the developing countries the macroeconomic impact of an energy price increase will occur in large part through indirect routes—for example, through changes in the constraints on the ability of a country to borrow the funds needed to import essential capital goods. At this time we cannot accurately predict just how these constraints will change.

Even if the LDC's enjoy rapid economic growth over the next two decades, this does not mean that their use of energy will increase rapidly. It is likely that in order to maintain high growth rates (and in particular to limit external indebtedness so that essential capital goods can be imported) imports of oil and other fuels will have to be restricted. Thus either by using taxes to raise retail prices (as many developing countries have already done) or simply by rationing limited supplies, energy consumption could be suppressed. It is likely that this will be the case to at least some extent, so that the developing countries' share of world energy consumption may not change very much.

**Table 7.1**
Demand estimates for Greece, Spain and Turkey

| Dependent variable | Price | GDP | Lagged dependent | Constants Greece | Spain | Turkey |
|---|---|---|---|---|---|---|
| 1. Light fuel oil | 0.385<br>(1.42) | 0.676<br>(2.29) | 0.760<br>(12.50) | −13.7<br>(−1.78) | −14.5<br>(−1.83) | −11.8<br>(−1.75) |
| 2. Light fuel oil | — | 0.653<br>(2.86) | 0.796<br>(12.0) | −5.34<br>(−2.51) | −6.09<br>(−2.61) | −4.10<br>(−2.40) |
| 3. Heavy fuel oil | −2.25<br>(−7.74) | 1.52<br>(5.97) | — | 1.16<br>(0.18) | 10.4<br>(1.33) | 10.9<br>(1.66) |
| 4. Heavy fuel oil | −1.01<br>(−3.70) | 0.365<br>(1.79) | 0.651<br>(7.55) | 6.11<br>(2.05) | 10.7<br>(2.54) | 9.78<br>(2.72) |
| 5. Motor gasoline | −0.159<br>(−1.54) | 0.744<br>(6.04) | 0.616<br>(8.75) | −2.99<br>(−1.67) | −3.37<br>(−1.76) | −2.03<br>(−1.31) |
| 6. Motor gasoline | −0.333<br>(−2.09) | 1.72<br>(20.90) | — | −6.80<br>(−2.49) | −7.71<br>(−2.67) | −4.55<br>(−1.91) |
| 7. Diesel fuel | — | 0.366<br>(1.98) | 0.761<br>(8.80) | −2.03<br>(−1.47) | −2.23<br>(−1.48) | −1.47<br>(−1.37) |
| 8. Kerosene | 0.163<br>(1.24) | −0.089<br>(−1.18) | 0.764<br>(10.1) | −0.055<br>(−0.03) | −0.143<br>(−0.06) | 0.006<br>(0.003) |
| 9. Kerosene | — | −0.177<br>(−3.93) | 0.787<br>(11.4) | 2.81<br>(4.24) | 2.88<br>(4.28) | 2.52<br>(4.26) |
| 10. Other petroleum products | −0.219<br>(−0.94) | 0.699<br>(1.58) | 0.570<br>(4.94) | 4.45<br>(0.32) | 4.99<br>(0.27) | 3.87<br>(0.32) |

[a] Price of electricity for other petroleum products.
[b] Time required to reach one-half of total adjustment in demand = log (0.5)/log $\lambda$.

| Price energy[a] | Temperature | Long-run elasticities Price | GDP | Median lag[b] | $R^2$ | $F$ |
|---|---|---|---|---|---|---|
| 0.108 (0.68) | 0.648 (0.97) | 1.60 | 2.81 | 2.53 | 0.973 | 181 |
| — | — | — | 3.20 | 3.04 | 0.981 | 288 |
| 0.841 (2.02) | — | −2.25 | 1.52 | — | 0.937 | 153 |
| — | — | −2.89 | 1.05 | 1.61 | 0.968 | 309 |
| — | — | −0.414 | 1.94 | 1.43 | 0.982 | 571 |
| — | — | −0.33 | 1.72 | — | 0.956 | 283 |
| — | — | — | 1.53 | 2.54 | 0.984 | 422 |
| — | — | 0.69 | −0.38 | 2.57 | 0.914 | 123 |
| — | — | — | −0.83 | 2.89 | 0.901 | 108 |
| −.434 (−0.62) | — | −0.51 | 1.62 | 1.23 | 0.966 | 215 |

**Table 7.2**
Demand estimates for Brazil and Mexico

| Dependent variable | Price | GDP | Lagged dependent | Constants | |
|---|---|---|---|---|---|
| | | | | Brazil | Mexico |
| 1. Fuel oils | −0.84 | 0.126 | 0.777 | 1.69 | 0.22 |
| | (−1.87) | (1.53) | (5.20) | (3.83) | (3.21) |
| 2. Gasolines | −0.118 | 0.260 | 0.787 | 1.48 | 0.603 |
| | (−1.72) | (2.41) | (6.21) | (2.35) | (1.13) |
| 3. Kerosene | −0.129 | 0.100 | 0.349 | 3.53 | 4.47 |
| | (−2.14) | (1.44) | (5.09) | (5.09) | (2.29) |
| 4. LPG | −0.762 | 1.72 | — | 4.97 | −0.81 |
| | (−3.45) | (6.21) | | (2.87) | (−2.41) |

| Long-run elasticities | | Median | | |
|---|---|---|---|---|
| Price | GDP | lag | $R^2$ | $F$ |
| −0.38 | 0.57 | 2.75 | 0.970 | 232 |
| −0.55 | 1.22 | 2.89 | 0.986 | 482 |
| −0.20 | 0.15 | 0.66 | 0.992 | 912 |
| −0.76 | 1.72 | — | 0.931 | 108 |

**Table 7.3**
Demand estimates for the industrialized countries[a]

| Dependent variable | Price | GDP | Lagged dependent | Other fuel prices | | |
|---|---|---|---|---|---|---|
| | | | | Coal | Electricity | Gas |
| A. U.S. and Canada | | | | | | |
| 1. Light fuel oil | −0.379 (−2.48) | 0.268 (2.39) | 0.561 (4.79) | — | — | — |
| 2. Light fuel oil | −0.279 (−1.82) | 0.224 (2.05) | 0.640 (5.44) | — | — | — |
| 3. Heavy fuel oil | −0.606 (−3.08) | 1.57 (3.62) | — | 1.42 (4.19) | 1.76 (2.75) | −0.024 (−0.06) |
| B. Europe | | | | | | |
| 4. Light fuel oil | −0.171 (−2.46) | 0.157 (1.67) | 0.856 (36.9) | — | — | — |
| 5. Light fuel oil | −0.127 (−1.86) | 0.220 (2.33) | 0.857 (38.3) | — | — | — |
| 6. Heavy fuel oil | −0.109 (−1.96) | 0.004 (0.04) | 0.890 (28.9) | −0.106 (−1.66) | 0.055 (0.69) | 0.074 (1.87) |
| 7. Heavy fuel oil | −0.101 (−2.06) | −0.015 (−0.17) | 0.879 (28.5) | — | — | — |

[a] Constant terms not included in table.

| Temperature | Long-run elasticities Price | GDP | Median lag | $R^2$ | $F$ |
|---|---|---|---|---|---|
| — | −0.86 | 0.61 | 1.20 | 0.970 | 256 |
| −0.377 (−2.09) | −0.78 | 0.62 | 1.55 | 0.973 | 228 |
| — | −0.61 | 1.57 | — | 0.943 | 180 |
| — | −1.19 | 1.09 | 4.46 | 0.988 | 1295 |
| −0.690 (−4.75) | −0.89 | 1.54 | 4.49 | 0.989 | 1246 |
| — | −0.99 | 0.04 | 5.95 | 0.961 | 304 |
| — | −0.84 | −0.12 | 5.37 | 0.959 | 361 |

# 8 Higher Energy Prices and Economic Growth

In chapter 1 we noted that the relationship between energy use and economic growth runs in both directions. Clearly changes in GNP are an important determinant of changes in energy demand (although a 1 percent increase in GNP might result in less than or more than a 1 percent increase in energy demand, depending on the particular sector of use and on the particular country). Now we wish to examine the extent to which GNP growth is dependent upon the use of energy, and in particular the extent to which an increase in the price of energy tends to reduce the potential productive capacity of the economy (in addition to reducing energy use).

Whenever a particular factor of production—in this case energy—becomes more scarce, that is, more costly, this necessarily reduces the economy's production possibilities. Thus, even if the proper monetary and fiscal policy were pursued so that the economy could operate close to its full-capacity level, an increase in the price of energy will mean a lower GNP. The question, of course, is, To what extent are higher energy prices likely to depress the level or future growth rate of output? In this chapter we deal with this question qualitatively and discuss how the answer depends on the various elasticities estimated earlier in this study.[1]

Shortages of energy occurring, for example, as a result of government-imposed controls on prices can force a production cutback in some industries and create bottlenecks in the overall flow of intermediate goods. Shortages can therefore have a much greater impact on economic output and employment than even major increases in energy prices, particularly in the short term. Analyzing the macroeconomic impact of an energy shortage, however, requires a detailed model of the interindustry structure of the economy, as well as information on the allocation of the shortage. Our concern here will therefore be limited to the impact of higher energy prices.

As we will see, a good case can be made that the sharp increases in energy prices that occurred in 1973 to 1974 indeed had a significant dampening effect on the output of the industrialized countries, although for many countries the macroeconomic adjustment to these price increases may have already largely occurred. Now, however,

1. For a more detailed attempt at a numerical analysis of this problem, see the recent work by Hogan (86) and Hogan and Manne (87).

we face another important question, namely, To what extent are energy prices likely to rise in the future? Forecasting the world price of energy, which to a certain extent will be tied to the behavior of the OPEC cartel, is not a major objective of this book, but we will nonetheless speculate on what we believe to be a most likely scenario. We will also discuss the question of whether energy supplies are likely to be available in the future at whatever price happens to prevail, or whether sudden shortages are more probable. Finally, we will offer some comments on the implications of this analysis for energy policy.

## 8.1 Energy and the Macroeconomy

The way in which the use and price of energy affect macroeconomic growth is determined largely by the role of energy in the structure of industrial production. It is in the industrial sector, where energy is an intermediate input to production, that an increase in its price will affect the production possibilities of the economy and thus potential output. Of course, increases in the price of energy resulting from growing domestic scarcity can also reduce the GNP contribution of the energy sector itself, and as the energy that is consumed directly (as in the residential sector) becomes more expensive, this reduces the total consumption possibilities of the economy (as measured by the physical quantities of goods consumed). Furthermore, in addition to its impact on real national income and national product, increases in the price of energy will increase the aggregate price index. Although the inflationary impact of higher energy prices is important and of concern to policy makers, our focus here will be the impact on the real productive capacity of the economy.

It is important to point out that the rate at which energy prices rise will be an important determinant of the magnitude of the macroeconomic impact, and that the impact in the short run will differ from that over the longer term. In the short run producers cannot respond to a sharp increase in energy prices by changing their production technologies or shifting to more energy-efficient machines. While the dynamics of the adjustment to higher energy prices is interesting, its analysis requires empirical tools that have not been constructed as part of this study. Our empirical findings on the structure of industrial energy demand pertain largely to the long run, and we will be interested in the implications of those findings for the long-run impact of higher energy prices.

In order to examine the relationship between energy prices and economic output, let us assume for simplicity that all energy is used as an intermediate input to the output of final goods and services. When the price of energy increases, the price (cost) of this output is pushed up (by an amount to be determined), and to the extent that other factors of production are not substituted for energy, the payments to these factors remain the same. In effect the real output (national income) available to these factors has decreased since greater resources must now be spent on the more expensive intermediate input. If, for example, all energy was imported, then increased resources (that is, increased capital and labor) would have to be allocated to the generation of the additional exports needed to offset the higher value of the imports. If the energy was produced domestically and became more expensive because of the depletion of proved and potential reserves, increased resources would be needed to explore for and extract a given amount of energy. In either case fewer resources will now be used to produce a smaller output of final goods and services available for domestic use.

We can thus use an aggregate cost function to measure the impact of an increase in the price of energy on real output—this impact is simply equal to the resulting increase in the cost of a fixed quantity of output. We must therefore examine the percentage increase in the cost of output (with the level of output fixed) resulting from a 1 percent increase in the price of energy—such that the energy price elasticity of the cost of output $\eta^*_{CE} = d\log C/d\log P_E$.

We are interested in the total elasticity of output cost, that is, an elasticity that accounts for changes in the prices and uses of other factors. Note that this is not the same as the partial elasticity derived in chapter 2, for which the prices and quantities of other factors are assumed fixed. To obtain this elasticity we expand the total derivative:

$$\frac{d\log C}{d\log P_E} = \frac{\partial\log C}{\partial\log P_E} + \frac{\partial\log C}{\partial\log P_K}\frac{\partial\log P_K}{\partial\log K}\frac{\partial\log K}{\partial\log P_E} + \frac{\partial\log C}{\partial\log P_L}\frac{\partial\log P_L}{\partial\log L}\frac{\partial\log L}{\partial\log P_E}. \tag{8.1}$$

We can now use the translog cost function of equation (2.39) and the translog share equations (2.40) to obtain the partial derivatives:

$$\frac{d\log C}{d\log P_E} = S_E + S_K\eta_{KE}\frac{\partial\log P_K}{\partial\log K} + S_L\eta_{LE}\frac{\partial\log P_L}{\partial\log L}. \tag{8.2}$$

For a single firm we can assume $\partial\log P_K/\partial\log K = \partial\log P_L/\partial\log L = 0$, so that $\eta^*_{CE} = S_E$, as in equation (2.55). While we would expect the supply functions for capital and labor that are faced by an individual firm to be infinitely elastic, this assumption is not reasonable for the economy as a whole. If factor markets are competitive and aggregate production exhibits decreasing returns to each factor, the factor prices $P_K$ and $P_L$ will fall as the factor quantities $K$ and $L$ increase.

If energy is substitutable with both capital and labor, that is, the cross-price elasticities $\eta_{KE}$ and $\eta_{LE}$ are both positive, an increase in the price of energy will result in an increase in the use of both capital and labor, and a decrease in the price of each. Observe from equation (8.2) that this will reduce the elasticity of the cost of output, such that this elasticity will now be smaller than the cost share of energy. It is difficult to determine just what value this elasticity will now have, as it depends on the extent to which factor prices change as factor quantities change, which in turn depends on the structure of factor markets.

Note that in addition to increasing the cost of output, an increase in the price of energy will in general also change the allocation of this cost to all of the factors, so that the relative shares of (the reduced) real product received by capital and labor may change. Using the translog cost shares of equation (2.40), observe that 1 percent increase in the price of factor $j$ results in a change in the share of factor $i$ of $dS_i/d\log P_j = \gamma_{ij}$. Using our estimates of the long-run cost function, we see from column 3 of table 5.7 that $\gamma_{EL} = -0.001$ and $\gamma_{EK} = -0.0056$. Thus according to our estimated cost function there is an increase in the share of labor relative to that of capital. Another way of putting this is that as the price of energy rises, there is simply more substitution of labor than there is of capital.

Now suppose that $\alpha_{KE}$, the elasticity of substitution between capital and energy, is negative, as suggested by Berndt and Wood (22) and by other studies (60, 62, 114). (Recall from chapter 5 our argument that a negative elasticity of substitution here probably reflects the short run. The short-run values of this and related elasticities are extremely relevant, however, if we are interested in determining the impact over a five-year period of a sharp and sudden increase in energy prices.) In this case reduced energy use will be accompanied by reduced use of capital, but a larger increase in the use of labor. As long as the potential supply of labor is available so that employment can indeed expand, this provides no particular problem (and over the long run may in fact lead to lower levels of structural unemployment). In fact, the elasticity of the cost of output will again be below the cost share of energy, since the negative cross-price elasticity between capital and energy must be accompanied by a positive cross-price elasticity between labor and energy that is larger in magnitude.

Suppose, on the other hand, that the supply of labor is fixed. In this case the percentage decrease in real output can be even larger than the cost share of energy $S_E$. In such a case an increase in the price of energy will be followed by a reduction in the use of both energy and capital, but not increase in labor. Here $\eta_{LE}$ is constrained to be 0, and

$$\frac{d\log C}{d\log P_E} = S_E + S_K \eta_{KE} \frac{\partial \log P_K}{\partial \log K} . \tag{8.3}$$

Note that this is larger than $S_E$ if $\eta_{KE}$ is negative (that is, energy and capital are complements).[2]

What happens to the distribution of output in this case? Observe that, although the level of output will fall, labor's share is likely to increase since with an increased demand for labor but a fixed quantity the wage rate should rise. Thus the net impact of an increase

2. This is the basis on which Hogan (86) and Hogan and Manne (87) show a magnification of the impact of higher energy prices on output when the elasticity of substitution between capital and energy is negative. Another way to view the problem is to assume that the input of capital falls to maintain a constant rate of return, while the input of labor is fixed. Hogan and Manne do this and show that for an elasticity of substitution of 0.3 between energy and a capital-labor aggregate, a 50 percent decrease in the use of energy results in a 4 percent drop in GNP with an increase in capital, but an 11 percent drop when the quantity of capital falls to maintain the rate of return.

in the price of energy is likely to be a decrease in total output with an elasticity greater than $S_E$, a decrease in profits (with a higher profit rate since the quantity of capital is lower), and a decrease in wages with an elasticity less than $S_E$.

We should also point out that the macroeconomic impact of an increase in the price of energy will also depend on the role of the energy-producing sector in the economy. Suppose, for example, that all energy consumed is produced domestically. An increase in the price of energy might result in an increase in production costs (due, for example, to depletion). In this case greater resources (that is, more capital and labor) are needed to produce the same amount of energy, and less resources are available for the production of other final goods and services, so that the GNP falls accordingly. As we explained earlier, the effect here is the same as if all energy were imported, in which case a higher price means an increase in import expenditures that (in the long run) must be matched by an export increase, which in turn requires an increased allocation of domestic capital and labor to produce the goods that are to be traded for the more expensive energy.

On the other hand, an increase in the price of energy might occur suddenly (for example, through the action of a cartel, or through a sudden change in expectations about potential reserves, future prices, and so forth), and domestic producers of energy might be holding reserve inventories that suddenly increase in value. If the windfalls that result are used to expand the domestic capital stock, this could ameliorate the initial economic impact of the higher energy price. The effect, however, is likely to be small, particularly over the long term, so that with reasonable approximation we can view an increase in the price of energy as implying a direct increase in the resources needed to produce domestic energy, or, equivalently, an increase in the cost of imported energy that must (ultimately) be matched by the allocation of resources to the production of exports.

Let us now try to summarize what we can say about the macroeconomic impact of higher energy prices. Clearly this impact will depend critically on the elasticities of substitution for capital, labor, and energy, in other words, on the structure of production. However, it will also depend on the structure of labor and capital markets. If the supplies of capital and labor are unlimited, then a 10 percent increase in the price of energy can at most decrease real output by 10 percent times the cost share of energy (for example, if all factors must be used in fixed proportions), although the distribution of this lower output between capital and labor will depend on the relative changes of the prices of these factors. Note that if the elasticity of substitution between, say, capital and energy is negative, the elasticity of substitution between labor and energy must be positive and larger in magnitude, so that capital-energy complementarity results in greater substitution of labor, and a smaller reduction in real output than would be the case with fixed factor proportions.

A 10 percent increase in the price of energy can cause real output to fall by more than 10 percent times the share of energy only if the elasticity of substitution between energy and a second factor is negative, and the third factor is limited in supply. Severe capital-energy complementarity could thus indeed magnify the impact of an energy price increase, but only if employment cannot be expanded.

What should the impact of a doubling of energy prices be, then, for a country like the U.S.? In the short term capital and energy are probably indeed complementary inputs, and, if the economy were operating at full capacity, the supply of labor would be limited. In this case the initial fall in real potential output could equal or even exceed the cost share of energy (which is about 4 percent in the U.S.). If, on the other hand, the economy were operating below capacity (as the U.S. economy was in 1974), then the drop in output might be significantly less than the share of energy.

Our results in chapter 5, however, indicate that over the longer term energy and capital become substitutable, so that the decrease in potential output (below what it would have been otherwise at any particular point in time) might be significantly less than the share of energy. As can be seen from equation (8.2), just how much less will depend on the elasticities of substitution and on the price elasticities of demand for the other factors (namely, on the second derivatives of the production function, if factors are priced in proportion to their marginal products).[3] What is interesting here, however, is that

3. To get some idea of what the long-run drop in potential output might be, let us use the estimates that we obtained in chapter 5 for the cross-price

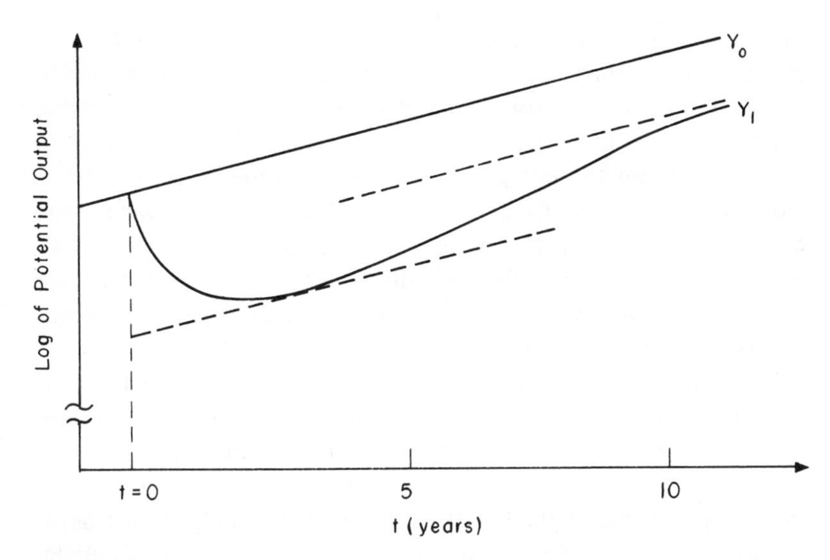


**Figure 8.1**
Impact of energy price increase on real potential output

initially the difference between potential output in the high energy price case and that in the low price case is larger than it is after a number of years have passed, so that in the interim period potential output might actually grow more rapidly (although from a lower level) than it would with a constant energy price. This is illustrated in figure 8.1. Of course, actual output may differ considerably from

---

elasticities $\eta_{KE}$ and $\eta_{LE}$, and let us assume that markets for labor and capital are competitive, so that $P_K = MP_K$ and $P_L = MP_L$. Now obtaining these marginal products from a translog production function with constant returns

$$\log F = \alpha_0 + \sum_i \alpha_i \log X_i + \sum_i \sum_j \gamma_{ij} \log X_i \log X_j,$$

we get $MP_i = (F/X_i)\, \partial \log F/\partial \log X_i = S_i F/X_i$. Then $\partial \log MP_i/\partial \log X_i = S_i - 1 + \gamma_{ii}$, and the decrease in output that follows a 100 percent increase in the price of energy is

$$S_E + \eta_{KE} S_K (S_K - 1 + \gamma_{KK}) + \eta_{LE} S_L (S_L - 1 + \gamma_{LL}).$$

Now using average values for $\gamma_{KK}$ and $\gamma_{LL}$ from the translog production functions estimated for U.S. manufacturing industries by Humphrey and Moroney (96), we get a drop in output of 0.034, as compared to the average cost share of energy of 0.04.

potential output.[4] If the economy is operating below capacity (as it more often is), then the supply of labor is likely to be less rigid in the short run, and, assuming capital-energy complementarity, the drop in actual output may initially be smaller. Furthermore, if in such a situation the energy price increase is met with an expansionary monetary and fiscal policy (which unfortunately was not the case in the U.S. and most other industrialized countries in 1974), actual output might not fall at all (although the potential growth rate over the next five years or so would not be as large as if energy prices had not increased).

The point here is that the observed changes in actual output (positive or negative) that follow a major increase in energy prices are the product not only of the direct reduction in potential output that we have just discussed but also the macroeconomic policies that happen to be applied at the time. Unfortunately such macroeconomic policies often tend (or at least in 1974 tended) to be contractionary, thus magnifying the recessionary impact of the price increase. The problem is quite simple: rising energy prices contribute directly to general inflation, and governments often try to fight this added inflation with contractionary macroeconomic policies.

This phenomenon can be illustrated by considering what happened when energy prices rose dramatically in 1974. As we would expect, the sharp increases in oil prices contributed directly to increased rates of inflation, particularly in 1974. In the U.S., for example, a good 3 to 4 percent of 1974s 11 percent inflation can be directly attributed to the increases in oil prices that occurred that year, while another 1.5 to 2 percent can be attributed to the higher food prices that resulted from increased demand for wheat and other food exports. This means that only 5 or 6 percent of the 11 percent inflation in that year was of the demand-pull variety that characterized inflation for the two decades prior to 1974, and that is responsive to conventional macroeconomic policy measures. And in Japan and many of the European countries, the inflationary impact of higher oil prices was even greater than in the U.S.

The problem occurred when most countries mistakenly re-

4. The measurement of potential output is itself no simple problem. For a recent example of the measurement of this quantity, see Perry (130).

sponded to this added inflation with strongly contractionary monetary and fiscal policies. The result was sharp recessions in 1975 in the U.S., Canada, and most of the European countries. Thus economic growth in 1974 through 1976 was lower in large part because of misguided macroeconomics policies—policies that might have worked well against the ordinary demand-pull inflation that governments had become used to, but that cannot possibly have an impact on the kind of exogenously generated inflation that was experienced in 1974. In fact, the economic impact of higher oil prices need not have been as serious as it was had the U.S. and the other industrialized countries responded with different monetary and fiscal policies.

For most of the industrialized countries the adjustment of potential output to the sharp increases in energy prices that occurred in 1973 to 1974 has largely already taken place. If our findings in chapter 5 regarding the long-run substitutability of energy with other factors are correct, and if further significant increases in real energy prices do not occur, then a return to something close to pre-1973 economic growth rates should soon be possible (see figure 8.1 again). In the U.S. and Canada, however, energy prices are (at the time of this writing) still well below world levels, largely because of government-imposed price controls. Assuming that energy prices in these countries will ultimately rise to the world level, a corresponding adjustment in potential output will be necessary. That adjustment will be modest, but certainly noticeable. In the U.S., for example, energy prices must rise an average of some 30 percent to reach the world level. Assuming no factor substitutability in the short run, this would imply a drop in potential output of just over 1 percent. Thus, assuming the adjustments in energy prices occur slowly over two to four years, annual growth rates should be lower over the period by some one-fourth to one-half percent.

What about increases in the price of energy in the longer term? We will argue shortly that over the next two decades energy prices will most likely rise only slowly, perhaps a few percent per year in real terms. At the very worst (that is, assuming no factor substitution possibilities), the annual economic growth rates of the industrialized countries would then be reduced by some 0.1 to 0.2 percent per year. Given that considerable factor substitution should be possible with energy prices that rise only gradually, the reduction in economic growth may hardly be noticeable.[5]

## 8.2 The Future Price and Availability of Energy

We have seen that if energy prices in the future rise only slowly in real terms, the growth rates of potential output in the industrialized countries can return to close to their pre-1973 values. It is therefore important to ask whether it is reasonable to expect only gradual increases in world energy prices, or whether we are more likely to experience one or more sharp jumps such as the one that occurred in 1973 to 1974. In addition, we must ask whether energy supplies are likely to be available even at higher prices, or whether world-wide shortages are likely to materialize. In the short run, the impact on economic output of a sudden shortage of energy supplies is likely to be much greater than the impact of a price increase.[6]

In the U.S., at least, the prospect of severe and long-lasting shortages of energy has been taken very seriously by government policy-makers, and has had an important role in the planning of long-range energy policy. Projections made by the CIA, as well as a number of private studies, predict that a crisis is likely to occur sometime during the 1980s as world energy demand exceeds supply, resulting in shortages of energy, rapidly rising prices, and economic contraction in all of the industrialized countries.[7] We will argue here that this view of the world is highly unrealistic, in that it ignores the impact of past and future changes in energy prices on

5. For a simulation study of the impact of higher energy prices on the OECD countries as a whole, which includes the effects of OPEC reinvestment in these countries, see Gunning, Osterrieth, and Waelbroeck (72). They find the impact on output to be relatively small, largely because of a prediction that the investment of foreign assets accumulated by the oil-producing countries will make a significant contribution to the GNP of OECD.

6. We will not discuss the measurement of the impact of a shortage of energy on GNP here. The interested reader might examine Klein (105) for one approach to the problem.

7. See, for example, the recent report of the Workshop on Alternative Energy Strategies (164).

energy supply and demand, and that the kind of crisis of concern to the CIA and others is very unlikely to occur.[8]

An exogeneous driving force behind the world price of energy and the evolution of world energy markets is the OPEC cartel. OPEC can, within limits, set the world price of oil, and thus it plays a major role in determining the world price of energy. In fact the crisis scenarios of the CIA and others are essentially based on the argument that OPEC will in the future either restrict its output because of capacity constraints that suddenly become binding, or else dramatically increase the price of oil as it did in 1973 to 1974. Let us begin then by examining OPEC and try to determine the kind of pricing behavior that the cartel is mostly likely to pursue.

Barring unforeseen political events, it seems most reasonable to expect OPEC to pursue whatever pricing policy is in its best economic interests. Of course, considerations other than economic ones might influence OPEC-pricing decisions, but economic interests have dominated in the past and are likely to dominate in the future, and provide the best basis for predicting oil prices. The ideal pricing policy from OPEC's point of view is one that maximizes the equity value of its oil resources—in other words, that maximizes the sum of all present and discounted future revenues from the sale of oil.

To determine its equity-maximizing price trajectory, OPEC must be concerned with three important issues. First, the demand for cartel oil is a residual demand, namely, the difference between the total (non-Communist) world demand for oil and the supply of oil from noncartel countries. This residual demand is particularly sensitive to price; we have seen in this study that the world demand for oil is reasonably price-elastic in the long run, and there is a growing body of statistical evidence which indicates that noncartel supplies of oil and other fuels may also be fairly responsive to price. Of course, the impact of a price change occurs only slowly, and this might suggest that OPEC could make a quick killing by steeply increasing the price and taking advantage of the fact that the demand for its oil would fall only after a delay of several years. Such a strategy indeed worked in 1973 when the world was accustomed to

8. The arguments that follow are developed in detail in Pindyck (138).

$4 per barrel oil, but it would not work now; it would increase current revenues for OPEC, but would diminish future revenues much more, with the net effect being a decrease in the equity value of OPEC's oil reserves.

A second consideration is the fact that OPEC oil resources are finite, and will eventually run out. This might suggest that OPEC should sharply increase price and cut production in order to conserve its resources for the future. But for the owner of an exhaustible resource, over-conservation is as bad as under-conservation—producing the resource too slowly simply reduces the total flow of current and future revenues that it can potentially yield. OPEC's problem—to choose just the right rate of resource use that balances revenue obtained today from current production with the (discounted) revenues that could be obtained from future production—is the problem facing any producer of an exhaustible resource, and its solution usually calls for smooth and gradual changes in price.

The third consideration is the fact that OPEC's membership consists of countries that have different objectives and operate under different constraints. Some OPEC members—which could be called "saver" countries—have less immediate need for cash, and therefore would tend to value future revenues more heavily in calculating the equity value of their resources. These countries, which include Saudi Arabia, Libya, and the Emirates, also have large reserves of oil, so that resource depletion is less of a constraint. Other OPEC members—the "spender" countries—have a greater immediate need for cash, and thus value current revenues more heavily. These countries, which include Iran, Venezuela, Algeria, Indonesia, Nigeria, and Ecuador, also have much smaller reserves of oil.

This division leads to a conflict over the optimal pricing strategy: spender countries would prefer higher prices initially than would saver countries. Much of the bargaining that goes on within OPEC involves reconciling this conflict. Although the outcome of this conflict cannot be predicted exactly, we have good reason to expect the interests of the saver countries to dominate, since it is these countries (and particularly Saudi Arabia) that have the greatest production capacity and at the same time absorb most of the necessary production cutbacks.[9]

9. In fact the interests of the saver countries dominated after the round of

In an earlier study (136) this author used a simple optimal control model to determine the implications of the considerations discussed above for OPEC's optimal oil-pricing policy. That model consisted of equations characterizing total world oil demand (and its response to income growth and price changes), non-OPEC oil supply (and its response to price changes as well as resource depletion), and resource depletion within OPEC. The solutions to the model yield optimal price trajectories that are quite robust with respect to assumptions about demand and supply elasticities and estimates of the size of the resource base, for example.

Solutions to the model indicated that OPEC's best price for 1977 was between \$12.50 and \$13.00—slightly below the actual posted price at the time. But more important, this price should grow in real terms by no more than 2 or 3 percent per year through 1990, and no more than 3 or 4 percent per year from 1990 to 2000. Prices rising more rapidly than this would simply cause a loss of revenue for OPEC, as the resulting loss in sales more than offsets the resulting increase from the higher price.

It is on the basis of these results—which in turn are based on the assumption that OPEC will pursue its economic objectives rationally—that we argue that world energy prices should rise only slowly in the future. Of course, we cannot be sure that OPEC will continue to pursue its economic objectives. Political events—such as another Mideast war, the destruction of oil-producing facilities by terrorists, or the overthrow of one of the key OPEC governments—might lead to a production cutback and, for at least a short period, a steep increase in price. (On the other hand, if such an event resulted in the dissolution of OPEC itself, it would lead to a decrease in price.) However, the possibility of such events only increases the forecast variance around our most likely scenario, and does not change the scenario itself.

Even if we are correct in our prediction that energy prices will rise only gradually, should we nonetheless expect to see serious world shortages of energy develop over the next two decades?[10]

---

OPEC meetings in May 1977 that resulted in a split price increase: Saudi Arabia simply increased her production slightly, so that the price increase was effectively limited.

10. The emphasis here is on world rather than national shortages. It is

Most predictions of world energy shortages really refer to oil shortages, and are based on the idea that sometime soon OPEC will not have the production capacity to meet the demands for its oil. Unfortunately these predictions usually ignore the fact that, as oil markets become tighter, prices rise and that the rising price itself limits demand, and encourages both non-OPEC energy production as well as OPEC capacity expansion.

To see why major shortages of oil are extremely unlikely, consider the fact that during 1977 and 1978 OPEC operated well below its production capacity, and in addition has considerable room for capacity expansion in the future. OPEC production in 1977, for example, was about 28 million barrels per day (mb/d), while capacity was over 35 mb/d. How is this capacity likely to change in the future? It is certainly quite likely that the capacity of such low-reserve oil-producing countries as Iran and Venezuela will decline during the 1980s and 1990s as reserves are depleted. It is very unlikely, however, that such high-reserve countries as Saudi Arabia, Kuwait, and the Emirates will keep their capacity fixed. In fact, Saudi Arabia and some of her high-reserve neighbors could as much as double their capacity, particularly if there is a revenue incentive to do so. And as long as these countries are interested in producing their oil at a rate that maximizes its equity value, they should increase their capacity, and exploit their oil resources at the optimum price. To restrict capacity and thereby over-conserve would be to waste the economic value of the resources.

Again all of this assumes that the OPEC countries base their oil production decisions on economic goals, and a breakdown in this assumption would permit a crisis to occur. As discussed earlier, unforeseen political events could indeed result in a sudden change in OPEC production and in the price of oil, and this could in turn have a disruptive impact on the industrialized economies. But such a crisis, which could occur as easily next year as in 1990, has not been of concern to those predicting shortages.

This view of world energy markets has implications for energy policy (and economic policy) in the U.S. and in the other industrial-

---

certainly conceivable that price controls imposed by the governments of individual countries could result in local shortages.

ized countries. First, the fear of an approaching crisis in which world energy demand suddenly exceeds supply, while perhaps useful for mobilizing public opinion behind some policy, should certainly not form the basis for policy formulation. On the other hand, a short-term, but sudden, production cutback occurring as an embargo or the result of political change in one or more OPEC countries, and causing a fast run-up in price perhaps combined with localized shortages, is a real possibility and is an important reason for pursuing domestic energy policies directed at limiting oil imports as well as maintaining strategic stockpiles of oil.

For some countries, notably the U.S., an unnecessary dependence on imported oil has been created through price controls on oil and natural gas together with related policies designed to maintain a low domestic price of energy. While the political benefits of regulating the price of energy to consumers is obvious, the costs of such policies are considerable. By reducing domestic energy production while raising consumption, these policies result in an evergrowing level of imports that must be paid for—directly or indirectly—by consumers.[11]

We have seen in this study that in all sectors of use the demand for energy is much more price elastic than had been thought to be the case. To the extent that limiting the growth of energy demand is a goal of energy policy, the price of energy itself may well be the most effective policy instrument available. This is particularly true

11. In the U.S., two principal policies have been used to maintain a low domestic price of energy. The first is the crude oil price controls-entitlements program which basically works by taxing the domestic production of oil (by holding its price below the refiner's price) and using the proceeds of the tax to subsidize imports (thereby reducing the cost of high-priced imported oil to the refiner). The second policy is the regulation of the wellhead price of natural gas which for many years has been held far below the world market level. Despite these low-price policies, the true cost of energy to the American consumer has been rising—as it must. The cost has come in part from growing imports of liquified natural gas (LNG) at prices double that of imported oil, from the taxes needed to subsidize noneconomical energy sources that would otherwise be unnecessary, and from the increased inflation resulting from the decreasing international value of the dollar brought about by rising oil imports. For a discussion of these aspects of U.S. energy policy, see Hall and Pindyck (75) and Pindyck (135).

in the U.S. and Canada, where prices have been held well below world market levels. If the supply of energy is also responsive to price (and there is a growing body of statistical evidence that it is), then the best policy for limiting imports would be one that deregulated domestic energy prices. As we have seen in this study, and as we have observed in some of the European countries which through taxes or other means have already increased their domestic energy prices, the use of energy is indeed responsive to higher prices.

# References

1. Adams, F. G., H. Graham, and J. M. Griffin. "Demand Elasticities for Gasoline: Another View." Discussion paper no. 279, Department of Economics, University of Pennsylvania, June 1974.

2. Adams, F. G., and J. M. Griffin. "Energy and Fuel Substitution Elasticities: Results from an International Cross Section Study." Unpublished manuscript, October 1974.

3. Adams, F. G., and P. Miovic. "On Relative Fuel Efficiency and the Output Elasticity of Energy Consumption in Western Europe." *Journal of Industrial Economics,* 22 (November 1968), pp. 41-56.

4. Allen, R. G. D. *Mathematical Analysis for Economists.* Macmillan, London, 1938.

5. Allen, R. G. D. *Index Numbers in Theory and Practice.* Aldine Publishing Company, Chicago, 1975.

6. Atkinson, S. E., and R. Halvorsen. "Interfuel Substitution in Steam Electric Power Generation." *Journal of Political Economy,* 84 (October 1976), pp. 959-978.

7. Belassa, B. "The Purchasing-Power Parity Doctrine: A Reappraisal." *Journal of Political Economy* (June 1964).

8. Barnett, H., and C. Morse. *Scarcity and Growth.* Johns Hopkins University Press, Baltimore, 1963.

9. Barten, A. P. "Estimating Demand Equations." *Econometrica,* 36 (April 1968), pp. 213-251.

10. Barten, A. P. "Maximum Likelihood Estimation of a Complete System of Demand Equations." *European Economic Review,* 1 (Fall 1969), pp. 7-73.

11. Baughman, M. L., and P. L. Joskow. "Interfuel Substitution in the Consumption of Energy in the United States." MIT Energy Laboratory working paper, Cambridge, Mass., May 1974.

12. Baxter, R. E., and R. Rees. "Analysis of the Industrial Demand for Electricity." *Economic Journal* (June 1968).

13. Belsley, D. A. "Estimation of Systems of Simultaneous Equations, and Computational Specifications of GREMLIN." *Annals of Economic and Social Measurement* (October 1974).

14. Berndt, E. R. "Canadian Energy Demand and Economic Growth." In *Oil in the Seventies*. Edited by G. C. Watkins and M. A. Walker. The Fraser Institute, Vancouver, 1977.

15. Berndt, E. R., and L. R. Christensen. "The Translog Function and the Substitution of Equipment, Structures, and Labor in U.S. Manufacturing 1929-68." *Journal of Econometrics,* 1 (March 1973), pp. 81-113.

16. Berndt, E. R., and L. R. Christensen. "The Internal Structure of Functional Relationships: Separability, Substitution, and Aggregation." *Review of Economic Studies,* vol. 40 (July 1973).

17. Berndt, E. R., M. N. Darrough, and W. E. Diewert. "Flexible Functional Forms and Expenditure Distributions: An Application to Canadian Consumer Demand Functions." *International Economic Review,* 18 (October 1977), pp. 651-676.

18. Berndt, E. R., M. A. Fuss, and L. Waverman. "Dynamic Models of the Industrial Demand for Energy." Electric Power Research Institute, Palo Alto, Calif. Technical report no. EA-580, November 1977.

19. Berndt, E. R., B. H. Hall, R. E. Hall, and J. A. Hausman. "Estimation and Inference in Nonlinear Structural Models." *Annals of Economic and Social Measurement,* 3 (October 1974), pp. 653-665.

20. Berndt, E. R., and N. E. Savin. "Estimation and Hypothesis Testing in Singular Equation Systems with Autoregressive Disturbances." *Econometrica,* 43 (September 1975), pp. 937-957.

21. Berndt, E. R., and G. C. Watkins. "Demand for Natural Gas: Residential and Commercial Markets in Ontario and British Columbia." *Canadian Journal of Economics,* 10 (February 1977), pp. 97-111.

22. Berndt, E. R., and W. Wood. "Technology, Prices, and the Derived Demand for Energy." *Review of Economics and Statistics,* 57 (August 1975), pp. 259-268.

23. Berndt, E. R., and D. O. Wood. "Consistent Projections of Energy Demand and Economic Growth: A Review of Issues and Empirical Studies." MIT Energy Laboratory working paper no. 77-024WP, Cambridge, Mass., June 1977.

24. Berndt, E. R., and D. O. Wood. "Engineering and Econometric Approaches to Industrial Energy Conservation and Capital Formation: A Reconciliation." MIT Energy Laboratory working paper no. 77-040WP, Cambridge, Mass., November 1977.

25. Brown, A., and A. Deaton. "Surveys in Applied Economics: Models of Consumer Behavior." *Economic Journal,* 82 (December 1972), pp. 1145-1236.

26. Burgess, D. F. "Duality Theory and Pitfalls in the Specification of Technologies." *Journal of Econometrics,* vol. 3 (1975).

27. Byron, P. "A Simple Method for Estimating Demand Systems under Separable Utility Assumptions." *Review of Economic Studies,* 37 (October 1970), pp. 261-274.

28. Carson, J. "A User's Guide to the MIT World Energy Demand Data Base." MIT Energy Laboratory Working Paper, 1978.

29. Chenery, H., and M. Syrquin. *Patterns of Development, 1950-1970.* Oxford University Press, London, 1975.

30. Chessire, J., and C. Buckley. "Energy Use in U.K. Industry." *Energy Policy,* 4 (September 1976), pp. 237-254.

31. Chow, G. "On the Computations of Full-Information Maximum Likelihood Estimates for Nonlinear Equation Systems." *Review of Economics and Statistics,* 55 (February 1973), pp. 104-109.

32. Christensen, L. R., D. W. Brazell, and D. Cummings. "Real Product, Real Factor Input, and Productivity in France, 1951-1973." Working Paper 7527. Social Systems Research Institute, University of Wisconsin at Madison, October 1975.

33. Christensen, L. R., and D. Cummings. "Real Product, Real Factor Input, and Productivity in Canada, 1947-1973." Working Paper 7532. Social Systems Research Institute, University of Wisconsin at Madison, October 1975.

34. Christensen, L. R., D. Cummings, and D. W. Jorgenson. "An International Comparison of Growth in Productivity, 1947-1973." Working Paper 7531. Social Systems Research Institute, University of Wisconsin at Madison, October 1975.

35. Christensen, L. R., D. Cummings, and B. Norton. "Real Product, Real Factor Input, and Productivity in Italy, 1952-1973." Working Paper 7528. Social Systems Research Institute, University of Wisconsin at Madison, October 1975.

36. Christensen, L. R., D. Cummings, and P. Schoech. "Real Product, Real Factor Input, and Productivity in the Netherlands, 1951-1973." Work-

ing Paper 7529. Social Systems Research Institute, University of Wisconsin at Madison, October 1975.

37. Christensen, L. R., D. Cummings, and K. Singleton. "Real Product, Real Factor Input, and Productivity in the United Kingdom, 1955-1973." Working Paper 7530. Social Systems Research Institute, University of Wisconsin at Madison, October 1975.

38. Christensen, L. R., and W. H. Greene. "Economies of Scale in U.S. Electric Power Generation." *Journal of Political Economy,* 84 (August 1976), pp. 655-676.

39. Christensen, L. R., and D. W. Jorgenson. "The Measurement of U.S. Real Capital Input, 1929-1967." *Review of Income and Wealth,* series 15 (1969), pp. 293-320.

40. Christensen, L. R., D. W. Jorgenson, and L. J. Lau. "Conjugate Duality and the Transcendental Logarithmic Function." *Econometrica,* 39 (July 1971), pp. 206-207.

41. Christensen, L. R., D. W. Jorgenson, and L. J. Lau. "Transcendental Logarithmic Production Frontiers." *Review of Economics and Statistics* 55 (February 1973), pp. 23-27.

42. Christensen, L. R., D. W. Jorgenson, and L. J. Lau. "Transcendental Logarithmic Utility Functions." *American Economic Review,* 65 (June 1975), 367-383.

43. Christensen, L. R. and M. E. Manser, "Estimating U.S. Consumer Preferences for Meat, 1947-71." Unpublished manuscript, February 1974.

44. Coen, R. M. "Effects of Tax Policy on Investment in Manufacturing." *American Economic Review,* vol. 58 (March 1968).

45. Committee on Nuclear and Alternative Energy Systems (CONAES), National Research Council. "U.S. Energy Demand: Some Low Energy Futures." *Science,* vol. 200 (April 14, 1978).

46. Corbo, V. and P. Meller. "The Translog Production Function: Some Evidence from Establishment Data." Unpublished manuscript, July 1977.

47. Cox, D. R. *Analysis of Binary Data.* Methuen and Co., London, 1970.

48. Darmstadter, J., J. Dunkerley, and J. Alterman. *How Industrial Societies Use Energy.* Johns Hopkins University Press, Baltimore 1977.

49. Deaton, A. S. "The Measurement of Income and Price Elasticities." *European Economic Review,* 6 (Fall 1975), pp. 261-273.

50. Denison, E. F. *Why Growth Rates Differ.* The Brookings Institution, Washington, D.C., 1967.

51. Denny, M., and M. Fuss. "The Use of Approximation Analysis to Test for Separability and the Existence of Consistent Aggregates." Working paper no. 7506. Institute for the Quantitative Analysis of Social and Economic Policy, University of Toronto, August 1975.

52. Diewert, W. E. "An Application of the Shephard Duality Theorem: A Generalized Leontief Production Function." *Journal of Political Economy,* 79 (May 1971), pp. 481-507.

53. Diewert, W. E. "Separability and a Generalization of the Cobb-Douglas Cost, Production, and Indirect Utility Functions." Working paper. University of British Columbia, 1973.

54. Diewert, W. E. "Homogeneous Weak Separability and Exact Index Numbers." Technical report no. 122. Institute for Mathematical Studies in the Social Sciences," Stanford University, Palo Alto, Calif. January 1974.

55. Doernberg, A. "Energy Use in Japan and the United States." Report no. BNL 50713. Brookhaven National Laboratory, Upton, N.Y. August 1977.

56. Domencich, T., and D. McFadden. *Urban Travel Demand: A Behavioral Analysis.* North-Holland Publishing Co., Amsterdam, 1975.

57. Edmonson, N. "Real Price and the Consumption of Mineral Energy in the United States, 1901-1968." *Journal of Industrial Economics,* vol. 23 (March 1975).

58. Eisner, M., and R. S. Pindyck. "A Generalized Approach to Estimation as Implemented in the TROLL/1 System." *Annals of Economic and Social Measurement,* 2 (January 1973), pp. 29-51.

59. Frisch, R. "A Complete Scheme for Computing All Direct and Cross Demand Elasticities in a Model with Many Sectors." *Econometrica,* 27 (April 1959), pp. 177-196.

60. Fuss, M. A. "The Demand for Energy in Canadian Manufacturing." *Journal of Econometrics,* 5 (1977), pp. 89-116.

61. Fuss, M. A. "The Derived Demand for Energy in the Presence of

Supply Constraints.'' Working paper no. 7714. Institute for Policy Analysis, University of Toronto, August 1977.

62. Fuss, M., and L. Waverman. ''The Demand for Energy in Canada.'' Working paper, Institute for Policy Analysis, University of Toronto, 1975.

63. Gallant, W. R. ''Seemingly Unrelated Nonlinear Regressions.'' *Journal of Econometrics* (February 1975).

64. Gilbert, M., and I. Kravis. *An International Comparison of National Products and the Purchasing Power of Currencies.* Organization for European Economic Cooperation, Paris, 1954.

65. Gilbert, M., and Associates. *Comparative National Products and Price Levels: A Study of Western Europe and the United States.* Paris, 1958.

66. Goldberger, A. S., and T. Gamaletsos. ''A Cross-Country Comparison of Consumer Expenditure Patterns.'' *European Economic Review,* 1 (Fall 1969), pp. 357-400.

67. Gregory, P., and J. M. Griffin. ''Secular and Cross-Section Industrialization Patterns: Some Further Evidence on the Kuznets-Chenery Controversy.'' *Review of Economics and Statistics* (August 1974).

68. Griffin, J. M. ''The Effects of Higher Prices on Electricity Consumption.'' *The Bell Journal of Economics and Management Science,* 5 (August 1974), pp. 515-539.

69. Griffin, J. M. ''An International Analysis of Demand Elasticities Between Fuel Types.'' Report to the National Science Foundation, 1977.

70. Griffin, J. M. ''Interfuel Substitution Possibilities: A Translog Application to Pooled Data.'' *International Economic Review,* 18 (October 1977), pp. 755-770.

71. Griffin, J. M., and P. R. Gregory. ''An Intercountry Translog Model of Energy Substitution Responses.'' *American Economic Review,* 66 (December 1976), pp. 845-857.

72. Gunning, J. W., M. Osterrieth, and J. Waelbroeck. ''The Price of Energy and Potential Growth of Developed Countries—An Attempt at Quantification.'' *European Economic Review,* 7 (Fall 1976), pp. 35-62.

73. Hall, R. E. ''Wages, Income, and Hours of Work in the U.S. Labor Force.'' In *Income Maintenance and Labor Supply.* Edited by G. Cain and H. Watts. Rand McNally, Chicago, 1973.

74. Hall, R. E., and D. W. Jorgenson. "Tax Policy and Investment Behavior," *American Economic Review,* 57 (June 1967), pp. 391-414.

75. Hall, R. E., and R. S. Pindyck. "The Conflicting Goals of National Energy Policy." *The Public Interest,* no. 47 (Spring 1977), pp. 3-15.

76. Halvorsen, R. "Residential Demand for Electric Energy." *Review of Economics and Statistics,* 57 (February 1975), pp. 12-18.

77. Halvorsen, R. "Energy Substitution in U.S. Manufacturing." Unpublished manuscript, 1976.

78. Halvorsen, R. "Disequilibrium Energy Demand in U.S. Manufacturing." Unpublished manuscript, September 1976.

79. Halvorsen, R., and J. Ford. "Substitution Among Energy, Capital, and Labor Inputs in U.S. Manufacturing." In *Advances in the Economics of Energy and Resources.* vol. 1. Edited by R. S. Pindyck. J. A. I. Press, Greenwich, Conn., 1978.

80. Hamilton, R. E. "World Energy Outlook to 1985." In *International Studies of the Demand for Energy.* Edited by W. D. Nordhaus. North-Holland Publishing Co., Amsterdam, 1977.

81. Hausman, J., and D. Wood. "Using the Share Ratio (Logit) Specification for Estimation of Price Elasticities in Demand Models." Unpublished manuscript, 1975.

82. Heide, R. J. "Log-Linear Models of Petroleum Product Demand: An International Study." MIT Energy Laboratory working paper, Cambridge, Mass., July 1978.

83. Heide, R. "An International Model of the Demand for Gasoline." MIT Sloan School of Management master's thesis, Cambridge, Mass., August 1978.

84. Hirst, E., W. Lin, and J. Cope. "An Engineering-Economic Model of Residential Energy Use." Technical report no. TM-5470, Oak Ridge National Laboratory, Tenn., July 1976.

85. Hoel, M. "A Note on the Estimation of the Elasticity of the Marginal Utility of Consumption." *European Economic Review,* 6 (Fall 1975), pp. 411-415.

86. Hogan, W. W. "Capital Energy Complementarity in Aggregate Energy-

Economic Analysis.'' Working paper. Institute for Energy Studies, Stanford University, Palo Alto, Calif., August 1977.

87. Hogan, W. W., and A. S. Manne. ''Energy-Economy Interactions: The Fable of the Elephant and the Rabbit?'' In *Advances in the Economics of Energy and Resources,* vol. 1. Edited by R. S. Pindyck. J. A. I. Press, Greenwich, Conn., 1978.

88. Hottel, H. C., and J. B. Howard. *New Energy Technology: Some Facts and Assessments*. The MIT Press, Cambridge, Mass., 1971.

89. Houthakker, H. S. ''An International Comparison of Household Expenditure Patterns, Commemorating the Centenary of Engel's Law.'' *Econometrica,* 25 (October 1957), pp. 532-551.

90. Houthakker, H. S. ''Additive Preferences.'' *Econometrica,* 28 (1960), pp. 244-257.

91. Houthakker, H. S., ''New Evidence on Demand Elasticities,'' *Econometrica,* 33 (April 1965), pp. 277-288.

92. Houthakker, H. S., and L. D. Taylor. *Consumer Demand in the United States: 1929-1970*. 2d ed. Harvard University Press, Cambridge, Mass., 1970.

93. Houthakker, H. S., P. K. Verleger, and D. P. Sheehan. ''Dynamic Demand Analyses for Gasoline and Residential Electricity.'' *American Journal of Agricultural Economics,* vol. 56 (May 1974).

94. Hudson, E. A., and D. W. Jorgenson. ''U.S. Energy Policy and Economic Growth, 1975-2000.'' *The Bell Journal of Economics and Management Science,* 5 (Autumn 1974), pp. 461-514.

95. Hulten, C. R. ''Divisia Index Numbers.'' *Econometrica,* 41 (November 1973), pp. 1017-1025.

96. Humphrey, D. B., and J. R. Moroney. ''Substitution Among Capital, Labor, and Natural Resource Products in American Manufacturing.'' *Journal of Political Economy,* 83 (February 1975), pp. 57-82.

97. Hymans, S. H. ''Consumer Durable Spending: Explanation and Prediction.'' *Brookings Papers on Economic Activity,* The Brookings Institution, Washington, D.C., No. 2. 1970, pp. 173-206.

98. Jorgenson, D. W. ''Consumer Demand for Energy.'' In *International*

*Studies of the Demand for Energy*. Edited by W. D. Nordhaus. North-Holland Publishing Co., Amsterdam, 1977.

99. Jorgenson, D. W., and Z. Griliches. "The Explanation of Productivity Change," *Review of Economic Studies,* 34 (July 1967), pp. 249-283.

100. Jorgenson, D. W., and Z. Griliches, "Issues In Growth Accounting: A Reply to Edward F. Denison." *Survey of Current Business,* 52 (May 1972), pp. 65-94.

101. Jorgenson, D. W., and L. J. Lau. "The Structure of Consumer Preferences." *Annals of Economic and Social Measurement,* 4 (Winter 1975), pp. 49-101.

102. Joskow, P. L., and M. L. Baughman. "The Future of the U.S. Nuclear Energy Industry," *Bell Journal of Economics,* 7 (Spring 1976), pp. 3-32.

103. Joskow, P. L., and F. S. Mishkin. "Electric Utility Fuel Choice Behavior in the United States." *International Economic Review,* 18 (October 1977), pp. 719-736.

104. Kesselman, J. R., S. H. Williamson, and E. R. Berndt. "Tax Credits for Employment Rather than Investment." *American Economic Review* (June 1977).

105. Klein, L. R. "Supply Constraints in Demand Oriented Systems: An Interpretation of the Oil Crisis." *Zeitschrift für Nationalökonomie,* 34 (1974), pp. 45-56.

106. Kloek, T., and H. Theil. "International Comparisons of Prices and Quantities Consumed." *Econometrica,* 33 (July 1965), pp. 535-556.

107. Kmenta, J., and R. Gilbert. "Small Sample Properties of Alternative Estimators of Seemingly Unrelated Regressions." *Journal of the American Statistical Association,* 63 (December 1968), pp. 1180-1200.

108. Kravis, I. B., Z. Kenessey, A. Heston, and R. Summers. *A System of International Comparisons of Gross Product and Purchasing Power.* Johns Hopkins University Press, Baltimore, 1975.

109. Lau, L. J. "Duality and the Structure of Utility Functions." *Journal of Economic Theory,* 1 (December 1969), pp. 374-396.

110. Lau, L. J., and S. Tamura. "Economies of Scale, Technical Progress, and the Nonhomothetic Leontief Production Function: An Application to

the Japanese Petrochemical Processing Industry." *Journal of Political Economy,* 80 (November 1972), pp. 1167-1187.

111. Liew, C. K. "Measuring the Substitutability of Energy Consumption." Unpublished manuscript, December 1974.

112. Lluch, C., and H. Powell. "International Comparisons of Expenditure Patterns." *European Economic Review,* 5 (Fall 1975), pp. 275-303.

113. MacAvoy, P. W., and R. S. Pindyck. *The Economics of the Natural Gas Shortage, 1960-1980.* North-Holland Publishing Co., Amsterdam, 1975.

114. Magnus, J. R. "Substitution Between Energy and Non-Energy Inputs in The Netherlands: 1950-1974." *International Economic Review* (forthcoming).

115. Manser, M. E. "The Translog Utility Function with Changing Tastes." Discussion paper no. 333. Economic Research Group, SUNY at Buffalo, N.Y. December 1974.

116. Manser, M. E. "Elasticities of Demand for Food: An Analysis Using Non-Additive Utility Functions Allowing for Habit Formation." Discussion paper no. 357. Economic Research Group, SUNY at Buffalo, N.Y., August 1975.

117. McFadden, D. "Conditional Logit Analysis of Qualitative Choice Behavior." In *Frontiers in Econometrics.* Edited by P. Zarembka. Academic Press, New York, 1973.

118. Mork, K. A. "The Inflationary Impact of Higher Energy Prices, 1973-75." MIT Energy Laboratory working paper, Cambridge, Mass., February 1978.

119. Moroney, J. R., and A. Toevs. "Factor Costs and Factor Use: An Analysis of Labor, Capital, and Natural Resource Inputs." *Southern Economic Journal,* vol. 44 (October 1977).

120. Moroney, J. R., and A. L. Toevs. "Input Prices, Substitution, and Product Inflation." In *Advances in the Economics of Energy and Resources,* vol. 1. Edited by R. S. Pindyck. J. A. I. Press, Greenwich, Conn., 1978.

121. Mount, T. D., L. D. Chapman, and T. J. Tyrrell. "Electricity Demand in the United States: An Econometric Analysis." Technical report. Oak Ridge National Laboratory, Tenn., June 1973.

122. Nelson, J. P. "The Demand for Space Heating Energy." *Review of Economics and Statistics,* 57 (November 1975), pp. 508–522.

123. Nordhaus, W. D. "The Demand for Energy: An International Perspective." In *International Studies of the Demand for Energy.* Edited by W. D. Nordhaus. North-Holland Publishing Co., Amsterdam, 1977.

124. Oberhofer, W., and J. Kmenta. "A General Procedure for Obtaining Maximum Likelihood Estimates in Generalized Regression Models." *Econometrica,* 42 (May 1974), pp. 579–590.

125. Organization for Economic Cooperation and Development. *Energy Prospects to 1985.* Paris, 1974.

126. Parikh, K. S. "Projecting and Planning India's Fuel Requirements." In *International Studies of the Demand for Energy.* Edited by W. D. Nordhaus. North-Holland Publishing Co., Amsterdam, 1977.

127. Park, R. E. "Prospects for Cable in the 100 Largest Television Markets." *Bell Journal of Economics and Management Science,* 3 (Spring 1972), pp. 130–150.

128. Parks, R. W. "Systems of Demand Equations: An Empirical Comparison of Alternative Functional Forms." *Econometrica,* 37 (October 1969), pp. 629–650.

129. Perloff, J. M., and M. L. Wachter. "A Production Function Non-Accelerating Inflation Approach to Potential Output: Is Measured Potential Output Too High?" University of Pennsylvania working paper, Philadelphia, April 1978.

130. Perry, G. L. "Potential Output and Productivity." *Brookings Papers on Economic Activity,* The Brookings Institution, Washington, D.C., no. 1. 1977.

131. Phlips, L. "A Dynamic Version of the Linear Expenditure Model." *Review of Economics and Statistics,* 54 (November 1972), pp. 450–458.

132. Phlips, L. *Applied Consumption Analysis.* North-Holland Publishing Co., Amsterdam, 1974.

133. Pindyck, R. S. "The Econometrics of U.S. Natural Gas and Oil Markets." In *Energy Modelling.* A Special *Energy Policy* publication. Surrey, England: IPE Science and Technology Press, 1974, pp. 124–133.

134. Pindyck, R. S. "International Comparisons of the Residential Demand

for Energy," MIT Energy Laboratory working paper no. 77-023-WP, Cambridge, Mass., August 1977.

135. Pindyck, R. S. "Prices and Shortages: Policy Options for the Natural Gas Industry." In *Options for U.S. Energy Policy*. Institute for Contemporary Studies, San Francisco, 1977.

136. Pindyck, R. S. "Gains to Producers from the Cartelization of Exhaustible Resources." *Review of Economics and Statistics,* 60 (May 1978), pp. 238-251.

137. Pindyck, R. S. "Higher Energy Prices and the Supply of Natural Gas." *Energy Systems and Policy*. New York: Crane, Russak and Co., Inc., 1978, pp. 177-209.

138. Pindyck, R. S. "OPEC's Threat to the West." *Foreign Policy,* 30 (Spring 1978), pp. 36-52.

139. Pindyck, R. S., "Interfuel Substitution and the Industrial Demand for Energy: An International Comparison." *Review of Economics and Statistics,* vol. 61 (May 1979).

140. Pindyck, R. S., and D. L. Rubinfeld. *Econometric Models and Economic Forecasts*. McGraw-Hill, New York, 1976.

141. Pollak, R. A. "Habit Formation and Dynamic Demand Functions." *Journal of Political Economy,* 78 (1970), pp. 745-763.

142. Pollak, R. A., and P. J. Wales. "Estimation of the Linear Expenditure System." *Econometrica,* 37 (1969), pp. 611-628.

143. Ramsey, J. B. "Limiting Functional Forms for Market Demand Curves." *Econometrica,* 40 (March 1972), pp. 327-341.

144. Ramsey, J., R. Rasche, and B. Allen. "An Analysis of the Private and Commercial Demand for Gasoline." *Review of Economics and Statistics,* 57 (November 1975), pp. 502-507.

145. Rødseth, A., and S. Strøm. "The Demand for Energy in Norwegian Households with Special Emphasis on the Demand for Electricity." Research memorandum. Institute of Economics, University of Oslo, April 1976.

146. Roy, R. *De l'Utilité—contribution à la théorie des choix*. Hermann, Paris, 1942.

147. Samuelson, P. A. "Analytical Notes on International Real Income Measures." *Economic Journal,* 84 (September 1974), pp. 595-608.

148. Sato, K., and T. Koizumi. "On the Elasticities of Substitution and Complementarity." *Oxford Economic Papers,* 25 (March 1973), pp. 44-56.

149. Shepard, R. W. *Cost and Production Functions.* Princeton University Press, Princeton, N.J., 1953.

150. Statistiches Bundesamt. "Internationaler Vergleich der Preise für die Lebenshaltung." *Preise Lohne Wirtschaftsrechsnungen,* Reihe 10, various years.

151. Strotz, R. H. "The Empirical Implications of a Utility Tree." *Econometrica,* 25 (1957), pp. 269-280.

152. Strout, A. M. "Energy and the Less Developed Countries: Needs for Additional Research." In *Changing Resource Problems of the Fourth World.* Edited by R. G. Ridker. Johns Hopkins University Press, Baltimore, 1976.

153. Strout, A. M. "Future Prospects for Nuclear Power in the Developing Countries." In *Advances in the Economics of Energy and Resources,* vol. 1. Edited by R. S. Pindyck. J. A. I. Press, Greenwich, Conn., 1978.

154. Taylor, L. D. "The Demand for Electricity: A Survey." *The Bell Journal of Economics,* 6 (Spring 1975), pp. 74-110.

155. Taylor, L. D. "The Demand for Energy: A Survey of Price and Income Elasticities." In *International Studies of the Demand for Energy.* Edited by W. D. Nordhaus. North-Holland Publishing Co., Amsterdam, 1977.

156. Taylor, L. D., and D. Weiserbs. "On the Estimation of Dynamic Demand Functions." *Review of Economics and Statistics,* 54 (November 1972), pp. 459-465.

157. Theil, H. "A Multinomial Extension of the Linear Logit Model." *International Economic Review,* 10 (October 1969), pp. 251-259.

158. Theil, H. *Theory and Measurement of Consumer Demand.* North-Holland Publishing Co., Amsterdam, 1975.

159. Tims, W. "The Developing Countries." In *Higher Oil Prices and the World Economy.* Edited by E. Fried and C. L. Schultze. The Brookings Institution, Washington, D.C., 1975.

160. Turvey, R., and A. R. Nobay. "On Measuring Energy Consumption," *Economic Journal,* 75 (December 1965), pp. 787-793.

161. U.K. Department of Energy. "Report of the Working Group on Energy Elasticities." Energy paper no. 17, February 1977.

162. Uzawa, H. "Production Functions with Constant Elasticities of Substitution." *Review of Economic Studies* (October 1962), pp. 291-299.

163. Wallace, T. D., and A. Hussain. "The Use of Error Components Model in Combining Cross-Section with Time Series Data." *Econometrica,* 37 (January 1969), pp. 55-72.

164. Workshop on Alternative Energy Strategies. *Energy: Global Prospects 1985-2000.* McGraw-Hill, New York, 1977.

165. Zellner, A. "An Efficient Method of Estimating Seemingly Unrelated Regressions and Tests for Aggregation Bias." *Journal of the American Statistical Association,* 57 (June 1962), pp. 348-368.

166. Zusman, P., E. Hochman, I. Luski, and E. Bezolel. "The Structure of Regional Demands for Energy." Unpublished technical report. Ben-Gurion University, Beer-Sheva, Israel.

# Index